高等院校通信与信息专业规划教材

通信工程专业英语

English for Telecommunication Engineering

刘金龙　张之光　龚育尔　钱　平　代小艳　编著

机械工业出版社

本书共12个单元，前6个单元介绍了通信发展的简史，并从“系统”的角度介绍了一些经典的和最新的通信系统，包括蜂窝移动通信系统、计算机网络通信系统、光纤通信系统、物联网、车联网、量子通信系统等；后6个单元介绍了通信工程领域的一些核心技术，涵盖交换技术、信息安全技术、复用和多址技术、调制解调技术、无线接入技术（WiFi、NFC）、数字信号处理技术等。每个单元又分为Text A和Text B两部分，A篇为该单元阅读和翻译教学的重点，B篇为A篇的补充和扩展，两者形式独立、内容相关。本书构思独特、选材新颖、内容丰富、语言规范、难易适中。每篇课文后面附有专业词汇、难点注释以及相应的练习，并且书后给出了每篇课文的参考译文，有助于学生的学习和理解。本书配有相应的课件，便于教学。

本书可作为高等院校通信工程及相关专业本科生“专业英语”课程的教材，也可供相关专业工程技术人员学习和参考。

图书在版编目（CIP）数据

通信工程专业英语/刘金龙等编著．—北京：机械工业出版社，2018.6（2024.1重印）

高等院校通信与信息专业规划教材

ISBN 978-7-111-60264-4

Ⅰ．①通…　Ⅱ．①刘…　Ⅲ．①通信－英语－高等学校－教材　Ⅳ．①H31

中国版本图书馆CIP数据核字（2018）第134477号

机械工业出版社（北京市百万庄大街22号　邮政编码100037）

策划编辑：李馨馨　责任编辑：李馨馨　杨　洋

责任校对：张　薇　封面设计：鞠　杨

责任印制：常天培

固安县铭成印刷有限公司印刷

2024年1月第1版第8次印刷

184mm×260mm·12.5印张·300千字

标准书号：ISBN 978-7-111-60264-4

定价：39.00元

凡购本书，如有缺页、倒页、脱页，由本社发行部调换

电话服务

服务咨询热线：010-88379833

读者购书热线：010-88379649

网络服务

机 工 官 网：www.cmpbook.com

机 工 官 博：weibo.com/cmp1952

教育服务网：www.cmpedu.com

金 书 网：www.golden-book.com

前　言

本书主要作为高等院校通信工程及相关专业本科生的“专业英语”课程教材使用，开课时间一般为第三或第四学年；也可供相关领域工程技术人员学习和参考。本书由通信专业博士和英语专业教师合作撰写，实现了专业和语言的有机结合，具有如下特点：

1. 构思独特

本书首先介绍了通信发展的简史以及代表人物——赫兹，然后从“系统”的角度对通信领域中一些经典的和最新的通信系统进行宏观介绍，最后对这些系统中采用的关键技术进行详细阐述。本书架构合理，内容全面，做到了“系统”和技术的有机结合。

2. 选材新颖

本书从大量的外文资料中遴选出了一些难度适中、语言严谨、能够展现当下最新的通信系统和核心技术的篇章作为素材，譬如物联网、车联网、量子通信、WiFi和NFC等。选材能够有效地拓展学生的视野，使其了解通信行业的最新发展动向。

3. 题材丰富

本书选编的24篇课文，题材涉及行业报告、人物传记、维基百科、新闻报道和技术文档等，能够让学生了解不同题材下的专业英语的风格和写作规范，有助于学生对专业英语在不同题材下呈现特点的把握。

此外，课文的编排也充分考虑到学生在先修大学英语时的学习模式，每篇课文后面附有专业词汇、难点注释以及相应的练习，书后给出了每篇课文的参考译文，有助于学生的学习和理解。本书配有相应的课件，便于教学。同时建议，课程的教授以每单元中的Text A为主，Text B作为选学内容，也可以在教师的指导下留给学生自学。

本书的策划、选材、翻译、校对与审核由刘金龙、张之光、龚育尔完成；Text A的课后词汇、注解、习题由代小艳编写；Text B的课后词汇、注解、习题与本书的课件由钱平、龚育尔编写。本书在编写过程中得到了南京航空航天大学电子信息工程学院雷磊副教授和南京邮电大学通信与信息工程学院刘旭副教授的悉心指导，在此对他们深表感谢。

限于编者水平，加之时间仓促，书中难免有疏漏和不妥之处，恳请读者使用后提出宝贵意见，以便今后能进一步改进和充实本书内容。

编　者

目　录

Unit 1

Text A

History of Telecommunication

The history of telecommunication began with the use of smoke signals and drums in Africa, the Americas and parts of Asia. In the 1790s, the first fixed semaphore systems emerged in Europe; however it was not until the 1830s that electrical telecommunication systems started to appear. This article details the history of telecommunication and the individuals who helped make telecommunication systems what they are today. The history of telecommunication is an important part of the larger history of communication.

1. Ancient systems and optical telegraphy

Early telecommunications included smoke signals and drums. Talking drums[1] were used by natives in Africa, New Guinea and South America, and smoke signals in North America and China. Contrary to what one might think, these systems were often used to do more than merely announce the presence of a military camp.

During the Middle Ages, chains of beacons were commonly used on hilltops as a means of relaying a signal. Beacon chains suffered the drawback that they could only pass a single bit of information, so the meaning of the message such as "the enemy has been sighted" had to be agreed upon in advance. One notable instance of their use was during the Spanish Armada, when a beacon chain relayed a signal from Plymouth to London that signaled the arrival of the Spanish warships.[2]

French engineer Claude Chappe began working on visual telegraphy in 1790, using pairs of "clocks" whose hands pointed at different symbols. These did not prove quite viable at long distances, and Chappe revised his model to use two sets of jointed wooden beams. Operators moved the beams using cranks and wires. He built his first telegraph line between Lille and Paris, followed by a line from Strasbourg to Paris (see Figure 1-1).

However, semaphore as a communication system suffered from the need for skilled operators and expensive towers often at intervals of only ten to thirty kilometers. As a result, the last commercial line was abandoned in 1880.

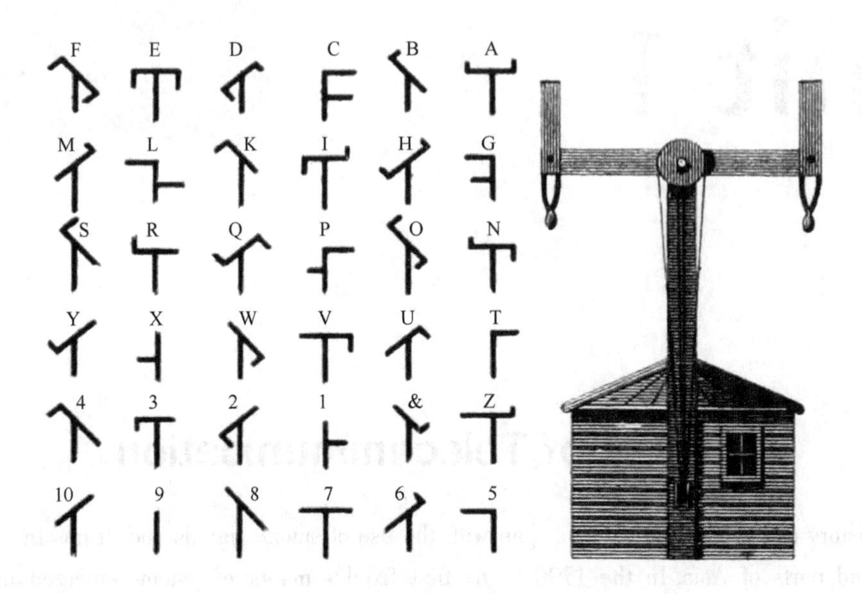

Figure 1-1

2. Electrical telegraph

Experiments on communication with electricity, initially unsuccessful, started in about 1726. Scientists including Laplace, Ampère, and Gauss were involved. The first working telegraph was built by Francis Ronalds in 1816 and used static electricity.

Charles Wheatstone and William Fothergill Cooke patented a five-needle, six-wire system, which entered commercial use in 1838. It used the deflection of needles to represent messages and started operating over twenty-one kilometers of the Great Western Railway on April 9, 1839. Both Wheatstone and Cooke viewed their device as "an improvement to the [existing] electromagnetic telegraph" not as a new device.

On the other side of the Atlantic Ocean, Samuel Morse developed a version of the electrical telegraph which he demonstrated on September 2, 1837. Alfred Vail saw this demonstration and joined Morse to develop the register—a telegraph terminal that integrated a logging device for recording messages to paper tape. This was demonstrated successfully over three miles (five kilometers) on January 6, 1838 and eventually over 40 miles (64 kilometers) between Washington, D. C. and Baltimore on May 24, 1844. The patented invention proved lucrative and by 1851 telegraph lines in the United States spanned over 20, 000 miles (32, 000 kilometers). Morse's most important technical contribution to this telegraph was the simple and highly efficient Morse Code, co-developed with Vail, which was an important advance over Wheatstone's more complicated and expensive system, and required just two wires. [3] The communications efficiency of the Morse Code preceded that of the Huffman code in digital communications by over 100 years, but Morse and Vail developed the code purely empirically, with shorter codes for more frequent letters.

3. Telephone

The electric telephone was invented in the 1870s; it was based on earlier work with harmonic (multi-signal) telegraphs. The first commercial telephone services were set up in 1878 and 1879 on both sides of the Atlantic in the cities of New Haven and London. Alexander Graham Bell held the master patent for the telephone that was needed for such services in both countries. All other patents for electric telephone devices and features flowed from this master patent. Credit for the invention of the electric telephone has been frequently disputed, and new controversies over the issue have arisen from time-to-time. As with other great inventions such as radio, television, the light bulb, and the digital computer, there were several inventors who did pioneering experimental work on voice transmission over a wire, who then improved on each other's ideas. [4] However, the key innovators were Alexander Graham Bell and Gardiner Greene Hubbard, who created the first telephone company, the Bell Telephone Company in the United States, which later evolved into American Telephone & Telegraph (AT&T), at times the world's largest phone company.

The first commercial telephone services were set up in 1878 and 1879 on both sides of the Atlantic in the cities of New Haven, Connecticut, and London, England. The technology grew quickly from this point, with inter-city lines being built and telephone exchanges in every major city of the United States by the mid-1880s. The First transcontinental telephone call occurred on January 25, 1915. Despite this, transatlantic voice communication remained impossible for customers until January 7, 1927 when a connection was established using radio.

4. Radio and television

Over several years starting in 1894 the Italian inventor Guglielmo Marconi built the first complete, commercially successful wireless telegraphy system based on airborne electromagnetic waves (radio transmission). In December 1901, Marconi established wireless communication between St. John's, Newfoundland and Poldhu, Cornwall (England), earning him a Nobel Prize in Physics (which he shared with Karl Braun). In 1900 Reginald Fessenden was able to wirelessly transmit a human voice.

For most of the twentieth century televisions used the cathode ray tube invented by Karl Braun. The first version of such a television to show promise was produced by Philo Farnsworth, who demonstrated crude silhouette images to his family in Idaho on September 7, 1927. Farnsworth's device would compete with the concurrent work of Kalman Tihanyi and Vladimir Zworykin. Though the execution of the device was not yet what everyone hoped it could be, it earned Farnsworth a small production company. In 1934, he gave the first public demonstration of the television at Philadelphia's Franklin Institute and opened his own broadcasting station.

After mid-century the spread of coaxial cable and microwave radio relay allowed television networks to spread across even large countries.

Television is not solely a technology limited to its basic and practical application. It functions both as an appliance, and also as a means for social story telling and message dissemination. It is a cultural tool that provides a communal experience of receiving information and experiencing fantasy. It acts as a "window to the world" by bridging audiences from all over through programming of stories, triumphs, and tragedies that are outside of personal experiences.

5. Satellite

The first U. S. satellite to relay communications was Project SCORE in 1958, which used a tape recorder to store and forward voice messages. It was used to send a Christmas greeting to the world from U. S. President Dwight D. Eisenhower.

Telstar was the first active, direct relay commercial communications satellite. Belonging to AT&T as part of a multi-national agreement among AT&T, Bell Telephone Laboratories, NASA, the British General Post Office, and the French National PTT (Post Office) to develop satellite communications, it was launched by NASA from Cape Canaveral on July 10, 1962, the first privately sponsored space launch. Relay 1 was launched on December 13, 1962, and became the first satellite to broadcast across the Pacific on November 22, 1963.

The first and historically most important application for communication satellites was in intercontinental long distance telephony. The fixed Public Switched Telephone Network relays telephone calls from land line telephones to an earth station, where they are then transmitted a receiving satellite dish via a geostationary satellite in Earth orbit. Improvements in submarine communications cables, through the use of fiber-optics, caused some decline in the use of satellites for fixed telephony in the late 20th century, but they still exclusively service remote islands such as Ascension Island, Saint Helena, Diego Garcia, and Easter Island, where no submarine cables are in service.

6. Computer networks and the Internet

On September 11, 1940, George Stibitz was able to transmit problems using Teletype 1 to his Complex Number Calculator in New York City and receive the computed results back at Dartmouth College in New Hampshire. This configuration of a centralized computer or mainframe with remote dumb terminals remained popular throughout the 1950s. However, it was not until the 1960s that researchers started to investigate packet switching—a technology that would allow chunks of data to be sent to different computers without first passing through a centralized mainframe. A four-node network emerged on December 5, 1969 among the University of California, Los Angeles, the Stanford Research Institute, the University of Utah and the University of California, Santa Barbara. This network would become ARPANET, which by 1981 would consist of 213 nodes. In June 1973, the first non-U. S. node was added to the network belonging to Norway's NORSAR project. This was shortly followed by a node in London.

ARPANET's development centered on the Request for Comment (RFC)[5] process and on

April 7, 1969, RFC 1 was published. This process is important because ARPANET would eventually merge with other networks to form the Internet and many of the protocols the Internet relies upon today were specified through this process. In September 1981, *RFC 791* introduced the Internet Protocol v4 (IPv4) and *RFC 793* introduced the Transmission Control Protocol (TCP)—thus creating the TCP/IP protocol that much of the Internet relies upon today. A more relaxed transport protocol that, unlike TCP, did not guarantee the orderly delivery of packets called the User Datagram Protocol (UDP) was submitted on 28 August 1980 as *RFC 768*. An e-mail protocol, SMTP, was introduced in August 1982 by *RFC 821* and http://1.0 a protocol that would make the hyperlinked Internet possible was introduced in May 1996 by *RFC 1945*.

Internet access became widespread late in the 20th century, using the old telephone and television networks.

New Words

semaphore	['seməfɔː (r)]	*n.* 臂板信号系统，(铁道) 臂板信号装置
optical	['ɒptɪkl]	*adj.* 视觉的，视力的；光学的
telegraphy	[tɪ'legrəfi]	*n.* 电信技术，超感
beacon	['biːk(ə)n]	*n.* 灯塔，信号浮标，烽火
relay	['riːleɪ]	*n.* 传递；继电器 *v.* 转播；分程传递
crank	[kræŋk]	*n.* [机] 曲柄 *v.* 转动曲柄移动；使弯曲
patent	['pætnt]	*n.* 专利 *v.* 获得……专利，给予……专利权
lucrative	['luːkrətɪv]	*adj.* 获利多的，赚钱的，合算的
span	[spæn]	*v.* 跨越时间或空间
empirically	[ɪm'pɪrɪkli]	*adv.* 以经验为主地
harmonic	[hɑː'mɒnɪk]	*adj.* 和声的，谐和的 *n.* [物] 谐波，和声
transmission	[træns'mɪʃn]	*n.* 播送，传送；传动装置
transcontinental	[ˌtrænzkɒntɪ'nentl]	*adj.* 横贯大陆的，大陆那边的
electromagnetic	[ɪˌlektrəʊmæg'netɪk]	*adj.* [物] 电磁的
cathode	['kæθəʊd]	*n.* [电] 阴极，负极
silhouette	[ˌsɪlʊ'et]	*n.* 轮廓，剪影；(事物的) 形状
coaxial	[kəʊ'æksɪəl]	*adj.* 同轴的，共轴的
communal	[kə'mjuːnl]	*adj.* 群体的，公民的，公共的
telephony	[tə'lefəni]	*n.* 电话学，电话，电话制造
geostationary	[ˌdʒiːəʊ'steɪʃnri]	*adj.* 与地球的相对位置不变的
configuration	[kənˌfɪgə'reɪʃn]	*n.* 布局，构造；配置
protocol	['prəʊtəkɒl]	*n.* (数据传递的) 协议

ARPANET: Advanced Research Project Agency (美国高级研究计划署)

TCP: Transmission Control Protocol (传输控制协议)

UDP: User Datagram Protocol (用户数据报协议)

Notes

[1] Talking drum（信息鼓）：西非的一种形状像沙漏的鼓，可以通过声音的调节来模仿人类语言的语调和韵律，从而传递不同的信息。

[2] One notable instance of their use was during the Spanish Armada, when a beacon chain relayed a signal from Plymouth to London that signaled the arrival of the Spanish warships.
Spanish Armada：无敌舰队是约有150艘以上的大战舰，3000余门大炮、数以万计士兵的强大海上舰队，最盛时舰队有千余艘舰船。这支舰队横行于地中海和大西洋，骄傲地自称为“无敌舰队”。
本句的意思是：关于使用信号灯的一个很典型的例子是在西班牙无敌舰队时期，当一连串的信号灯从普利茅斯传递到伦敦时则示意了西班牙战舰的到来。

[3] Morse's most important technical contribution to this telegraph was the simple and highly efficient Morse Code, co-developed with Vail, which was an important advance over Wheatstone's more complicated and expensive system, and required just two wires.
Morse code 莫尔斯码是一种时通时断的信号代码，通过不同的排列顺序来表达不同的英文字母、数字和标点符号。它发明于1837年，发明者有争议，是美国人塞缪尔·莫尔斯或者艾尔菲德·维尔。莫尔斯码是一种早期的数字化通信形式，但是它不同于现代只使用0和1两种状态的二进制代码，它的代码包括五种：点、划、点和划之间的停顿、每个字符间短的停顿（在点和划之间）、每个词之间中等的停顿以及句子之间长的停顿。
本句的意思是：莫尔斯对于电报最重要的贡献是他与韦尔一同发明的简单并且高效的莫尔斯码，仅需要两根电线即可实现，这是在惠斯通复杂且昂贵的系统之上一次巨大的进步。

[4] As with other great inventions such as radio, television, the light bulb, and the digital computer, there were several inventors who did pioneering experimental work on voice transmission over a wire, who then improved on each other's ideas.
as with 意为“正如，与……一样，就……来说”；句中两个 who 引导两个定语从句修饰先行词 inventors。
本句的意思是：正如其他伟大的发明，如收音机、电视、电灯、计算机，都有一些发明家进行电线传输声音的先驱试验，然后在彼此的想法上不断改进。

[5] RFC（Request for Comments）：是一系列以编号排定的文件。文件收集了有关互联网的相关信息，以及 UNIX 和互联网社区的软件文件。目前 RFC 文件是由 Internet Society（ISOC）赞助发行的。基本的互联网通信协议在 RFC 文件内都有详细说明。

Questions for Discussion

1. What's the main idea of this text?
2. What do you learn about the first commercial telephone?
3. How did Samuel Morse contribute to the development of telegraph?

Text B

Biography of Heinrich Hertz[1]

In a series of brilliant experiments Heinrich Hertz[1] discovered radio waves and established that James Clerk Maxwell's theory of electromagnetism is correct. Hertz also discovered the photoelectric effect[2], providing one of the first clues to the existence of the quantum world. The unit of frequency, the hertz, is named in his honor.

1. School

Aged six, Heinrich began at the Dr. Wichard Lange School in Hamburg. This was a private school for boys run by the famous educator Friedrich Wichard Lange. The school operated without religious influence; it used child-centered teaching methods, taking account of students' individual differences. It was also strict; the students were expected to work hard and compete with one another to be top of the class. Heinrich enjoyed his time at school, and indeed was top of his class. Unusually, Dr. Lange's school did not teach Greek and Latin—the classics—needed for university entry. The very young Heinrich had told his parents he wanted to become an engineer. When they looked for a school for him, they decided that Dr. Lange's alternative focus, which included the sciences, was the best option.

Heinrich's mother was especially passionate about his education. Realizing he had a natural talent for making things and for drawing, she arranged draftsmanship lessons for him on Sundays at a technical college. He started these aged 11.

2. Homeschool and Building Scientific Apparatus

Aged 15, Heinrich left Dr. Lange's school to be educated at home. He had decided that perhaps he would like to go to university after all. Now he received tutoring in Greek and Latin to prepare him for the exams.

He excelled at languages, a gift he seems to have inherited from his father.

Professor Redslob, a language specialist who gave Heinrich some tuition in Arabic, advised his father that Heinrich should become a student of oriental languages. Never before had he met anyone with greater natural talent.

Heinrich also began studying the sciences and mathematics at home, again with the help of a private tutor. He had a colossal appetite for hard work. His mother said, "When he sat with his books nothing could disturb him or draw him away from them".

Although he had left his normal school, he continued attending the technical college on Sunday mornings. In the evenings he worked with his hands. He learned to operate a lathe. He built models, and then began constructing increasingly sophisticated scientific apparatus, such as a spectroscope. He used this apparatus to do his own physics and chemistry experiments.

3. Becoming a Scientist

(1) Physics in Munich

In spring 1876, aged 19, he moved again, to Dresden, to study engineering. After only a few months he was drafted into the army for a year's compulsory service. After completing his army service, the 20-year-old Hertz moved to Munich to begin an engineering course in October 1877. A month later, after much internal anguish, he dropped out of the course. He had decided that above all else he wanted to become a physicist. Then he enrolled at the University of Munich, choosing courses in advanced mathematics and mechanics, experimental physics, and experimental chemistry. After a successful year at Munich he moved to the University of Berlin because it had better physics laboratories than Munich.

(2) Berlin, Helmholtz, and Recognition

In Berlin, aged 21, Hertz began working in the laboratories of the great physicist Hermann von Helmholtz.

Helmholtz must have recognized a rare talent in Hertz, immediately asking him to work on a problem whose solution he was particularly interested in. The problem was the subject of a fierce debate between Helmholtz and another physicist by the name of Wilhelm Weber.

The University of Berlin's Philosophy Department, with Helmholtz's encouragement, had offered a prize to anyone who could solve the problem: Does electricity move with inertia? Alternatively, we could frame the question in the form: Does electric current have mass? Or, as framed by Hertz: Does electric current have kinetic energy?

Hertz started work on the problem and quickly fell into a pleasant routine: Attending a lecture each morning in either analytical dynamics or electricity & magnetism, carrying out experiments in the laboratory until 4pm, then reading, calculating, and thinking in the evening.

He personally designed experiments which he thought would answer Helmholtz's question. He began to really enjoy himself, writing home:

"*I cannot tell you how much more satisfaction it gives me to gain knowledge for myself and for others directly from nature, rather than to be merely learning from others and myself alone.*"

4. The Prize

In August 1879, aged 22, Hertz won the prize—a gold medal. In a series of highly sensitive experiments he had demonstrated that if electric current has any mass at all, it must be incredibly small. We have to bear in mind that when Hertz carried out this work the electron—the carrier of electric current—had not even been discovered. J. J. Thomson's discovery was made in 1897, 18 years after Hertz's work.

Other physicists began to notice just how dazzling Hertz's work had been—the young student had put together experiments at the forefront of physics, personally modifying apparatus as needed. His practical skills, developed at home in the evenings, were proving to be priceless. His prize-winning work was published in the prestigious journal *Annalen der Physik*.

Recognizing the incredible talent he had in his laboratory, Helmholtz now asked Hertz to compete for a prize offered by the Berlin Academy: Verifying James Clerk Maxwell's theory of electromagnetism. Maxwell had stated in 1864 that light was an electromagnetic wave and that other types of electromagnetic wave could exist.

5. Doctor of Physics

Hertz declined this project; he believed the attempt, with no guarantee of success, would take several years of work. He was ambitious and wanted to publish new results quickly to establish his reputation.

Instead of working for the prize, he carried out a masterful three-month project on electromagnetic induction. He wrote this up as a thesis. In February 1880, at the age of 23, his thesis brought him the award of a doctorate in physics. Helmholtz quickly appointed him as an assistant professor. Later that year Hertz wrote:

"*I grow increasingly aware, and in more ways than expected, that I am at the center of my own field; and whether it be folly or wisdom, it is a very pleasant feeling.*"

Hertz stayed in Helmholtz's laboratory until 1883, during which time he published 15 papers in academic journals.

6. Mathematical Physics at Kiel

Hertz was a gifted experimental physicist, but competition to secure a lectureship at Berlin was high. Instead, with Helmholtz's support, Hertz became a lecturer in mathematical physics at the University of Kiel. This position, theoretical rather than experimental, extended his abilities. At Kiel he began to get to grips with Maxwell's equations, writing in his diary:

"*Hard at Maxwellian electromagnetism in the evening. Nothing but electromagnetism.*"

—HEINRICH HERTZ

Diary, May 1884

The result of Hertz's work was a highly regarded paper comparing Maxwell's electromagnetic theory with competing theories. He concluded that Maxwell's theory looked the most promising. In fact he reworked Maxwell's equations into a more convenient form.

He later wrote:

"*From the start, Maxwell's theory was the most elegant of all... the fundamental hypothesis of Maxwell's theory contradicted the usual views, and was not supported by evidence from decisive experiments.*"

—Heinrich Hertz

Diary, May 1884

7. The Discovery of Radio Waves

(1) Well-Equipped Laboratories and Attacking the Greatest Problem

In March 1885, desperate to return to experimental physics, Hertz moved to the University of

Karlsruhe. Aged 28, he had secured a full professorship. He was actually offered two other full professorships, a sign of his flourishing reputation. He chose Karlsruhe because it had the best laboratory facilities.

Wondering about which direction his research should take, his thoughts drifted to the prize work Helmholtz had failed to persuade him to do six years earlier: proving Maxwell's theory by experiment. Hertz decided that this mighty undertaking would be the focus of his research at Karlsruhe.

(2) A Spark That Changed Everything

After some months of experimental trials, the apparently unbreakable walls that had frustrated all attempts to prove Maxwell's theory began crumbling.

It started with a chance observation early in October 1886, when Hertz was showing students electric sparks. Hertz began thinking deeply about sparks and their effects in electric circuits. He began a series of experiments, generating sparks in different ways.

He discovered something amazing. Sparks were producing a regular electrical vibration within the electric wires they jumped between. The vibration moved back and forth more often every second than anything Hertz had ever encountered before in his electrical work.

He knew the vibration was made up of rapidly accelerating and decelerating electric charges. If Maxwell's theory were right, these charges would radiate electromagnetic waves which would pass through air just as light does.

8. Producing and Detecting Radio Waves

In November 1886 Hertz constructed the apparatus shown in Figure 1-2.

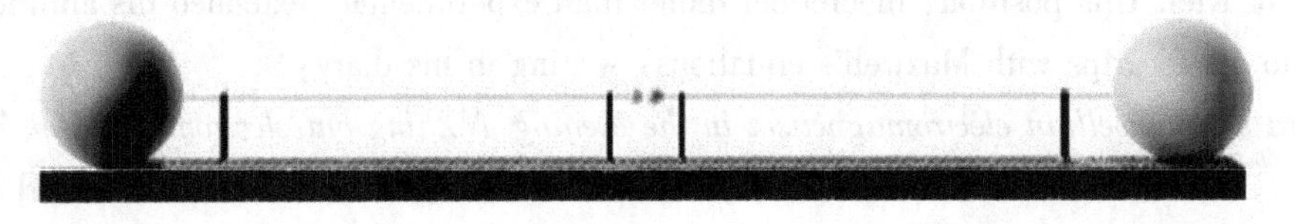

Figure 1-2

The Oscillator. At the ends are two hollow zinc spheres of diameter 30 cm. The spheres are each connected to copper wires which run Into the middle where there is a gap for sparks to jump between.

He applied high voltage A. C. electricity across the central spark-gap, creating sparks.

The sparks caused violent pulses of electric current within the copper wires. These pulses reverberated within the wires, surging back and forth at a rate of roughly 100 million per second.

As Maxwell had predicted, the oscillating electric charges produced electromagnetic waves—radio waves—which spread out through the air around the wires. Some of the waves reached a loop of copper wire 1. 5 meters away, producing surges of electric current within it. These surges caused sparks to jump across a spark-gap in the loop.

This was an experimental triumph. Hertz had produced and detected radio waves. He had passed electrical energy through the air from one device to another one located over a meter away. No connecting wires were needed.

"I do not think that wireless waves I have discovered will have any practical application."

Heinrich Hertz, 1890.

9. Taking It Further

Over the next three years, in a series of brilliant experiments, Hertz fully verified Maxwell's theory. He proved beyond doubt that his apparatus was producing electromagnetic waves, demonstrating that the energy radiating from his electrical oscillators could be reflected, refracted, produce interference patterns, and produce standing waves just like light.

Hertz's experiment's proved that radio waves and light waves were part of the same family, which today we call the electromagnetic spectrum. Strangely, though, Hertz did not appreciate the monumental practical importance of the electromagnetic waves he had produced.

This was because Hertz was one of the purest of pure scientists. He was interested only in designing experiments to entice nature to reveal its mysteries to him. Once he had achieved this, he would move on, leaving any practical applications for others to exploit.

The waves Hertz first generated in November 1886 quickly changed the world.

By 1896 Guglielmo Marconi[3] had applied for a patent for wireless communications. By 1901 he had transmitted a wireless signal across the Atlantic Ocean from Britain to Canada.

Hertz's discovery was the foundation stone for much of our modern communications technology. Radio, television, satellite communications, and mobile phones all rely on it. Even microwave ovens use electromagnetic waves: The waves penetrate the food, heating it quickly from the inside.

Our ability to detect radio waves has also transformed the science of astronomy. Radio astronomy has allowed us to "see" features we can't see in the visible part of the spectrum. And because lightning emits radio waves, we can even listen to lightning storms on Jupiter and Saturn. Scientists and non-scientists alike owe a lot to Heinrich Hertz.

New Words

electromagnetism	[ɪˌlektrəʊˈmægnɪtɪz(ə)m]	*n.* [电磁] 电磁学
quantum	[ˈkwɒntəm]	*n.* 量子论；额（特指定额、定量）
excel	[ɪk'sel; ek-]	*v.* 超过，擅长；（在某方面）胜过（或超过）别人
colossal	[kə'lɒs(ə)l]	*adj.* 巨大的；异常的，非常的
lathe	[leɪð]	*v.* 用车床加　*n.* 车床；机床
spectroscope	[ˈspektrəskəʊp]	*n.* [光] 分光镜
apparatus	[ˌæpəˈreɪtəs]	*n.* 装置，设备；仪器；器官
fierce	[fɪəs]	*adj.* 凶猛的；猛烈的；暴躁的
inertia	[ɪ'nɜːʃə]	*n.* [力] 惯性；惰性，迟钝；不活动
dazzle	[ˈdæz(ə)l]	*n.* 灿烂　*v.* 使……目眩；眼花缭乱，炫耀
grip	[grɪp]	*n.* 紧握　*v.* 紧握，夹紧；抓住

mighty	[ˈmaɪti]	*adj.* 有力的 *adv.* 很；非常 *n.* 有势力的人
crumble	[ˈkrʌmb(ə)l]	*v.* 崩溃；破碎，粉碎 *n.* 面包屑
vibration	[vaɪˈbreɪʃ(ə)n]	*n.* 振动；犹豫；心灵感应
radiate	[ˈreɪdieɪt]	*v.* 辐射，传播；辐射，从中心向各方伸展
oscillating	[ɒsɪˈleɪtɪŋ]	*adj.* ［物］振荡的
astronomy	[əˈstrɒnəmi]	*n.* 天文学

Notes

[1] Heinrich Hertz：海因里希·鲁道夫·赫兹（Heinrich Rudolf Hertz，1857—1894），德国物理学家，于1888年首先证实了电磁波的存在，对电磁学有很大的贡献。频率的国际单位制单位赫兹以他的名字命名。

[2] The Photoelectric Effect：光电效应是物理学中一个重要而神奇的现象。在高于某特定频率的电磁波照射下，某些物质内部的电子会被光子激发出来而形成电流，即光生电。光电现象由德国物理学家赫兹于1887年发现，而正确的解释为爱因斯坦所提出。在研究光电效应的过程中，物理学家对光子的量子性质有了更加深入的了解，这对波粒二象性概念的提出有重大影响。

[3] Guglielmo Marconi：古列尔莫·马可尼（Guglielmo Marconi，1874—1937），1874年4月25日生于博洛尼亚，意大利工程师，专门从事无线电设备的研制和改进；1909年诺贝尔物理学奖得主，被称为“无线电之父”。

Questions for Discussion

1. How did Heinrich learn Greek and Latin? Why?
2. What kind of routine did Hertz fall into at the University of Berlin?
3. How did he prove Maxwell's theory?

Unit 2

Text A

Mobile Wireless Overview

A fast-paced technological transition is occurring today in the world of internetworking. This transition is marked by the convergence of the telecommunications infrastructure with that of IP data networking to provide integrated voice, video, and data services.

As this transition progresses, the corresponding standards are continuing to evolve and many new standards are being developed to enable and accelerate this convergence of telecommunications and IP networking to mobilize the Internet and provide new multimedia services.

1. Introduction to Mobile Wireless Technology

The technologies related to wireless communication can be complex to differentiate. Wireless technology has been around for a while; however, there has been a relatively recent and rapid surge in the evolution of new wireless standards to support the convergence of voice, video and data communication. Much of this rapid evolution, or revolution, is a result of people seeking ubiquitous and immediate access to information and the assimilation of the Internet into business practices and for personal use. People "on the go" want their Internet access to move with them, so that their information is available at anytime, anywhere.

There are many factors that can be used to characterize wireless technologies:

- Spectrum, or the range of frequencies in which the network operates.
- Transmission speeds supported.
- Underlying transmission mechanism, such as frequency division multiple access (FDMA), time division multiple access (TDMA), or code division multiple access (CDMA).
- Architectural implementation, such as enterprise based (or in-building), fixed, or mobile.

In addition, the mobile wireless technologies, such as Global System for Mobile Communications (GSM), TDMA, CDMA are differentiated by a number of different factors, including some of the following:

- Control of the transmitted power;
- Radio resource management and channel allocation;
- Coding algorithms;
- Network topology and frequency reuse;
- Handoff mechanisms.

As suggested by its name, mobile wireless communication addresses those wireless technologies that support mobility of a subscriber, which provide seamless and real-time services without interruption. [1] Mobile wireless technologies support network access whether subscribers roam within or outside their home wireless coverage area.

2. Overview of Basic Network Elements Associated with Cellular Networks and Mobile Wireless

This section provides a brief introduction to a few of the basic network components associated with the existing telecommunications infrastructure. It specifically discusses the existing mobile wireless network infrastructure components for TDM-based wireless networks, some of which eventually will be replaced by new IP-based components.

In the early 1980s, support for mobile wireless communications was introduced using cellular networks, which were based on analog technologies such as AMPS[2]. Many of the telecommunications entities associated with cellular networks still play a vital role in today's wireless networks. As wireless communications technologies continue to progress and IP data networking is further integrated into the existing infrastructure, some of the functions of these entities might still exist within the network, but will be implemented in different and more effective ways.

The following network elements are part of a typical cellular telecommunications network:

- public switched telephone network (PSTN);
- mobile switching center (MSC);
- base station (BS);
- radio access network (RAN);
- home location register (HLR);
- visitor location register (VLR);
- authentication center (AUC). [3]

(1) Public Switched Telephone Network (PSTN)

The PSTN is the foundation and remains the predominant infrastructure that currently supports the connection of millions of subscribers worldwide. The PSTN has several thousands of miles of transmission infrastructure, including fixed land lines, microwave, and satellite links. After the introduction of cellular telephone systems in the early and mid-1980s, and with the rapid development of mobile wireless communication services, the PSTN still provides the fixed network support using the Signaling System Number 7 (SS7) protocol to carry control and signaling messages in a packet-switched environment.

(2) Mobile Switching Center (MSC)

The MSC, usually located at the Mobile Telephone Switching Office (MTSO), is part of the mobile wireless network infrastructure that provides the following services:

- Switches voice traffic from the wireless network to the PSTN if the call is a mobile-to-landline call, or it switches to another MSC within the wireless network if the call is a mobile-to-mobile call.
- Provides telephony switching services and controls calls between telephone and data systems.
- Provides the mobility functions for the network and serves as the hub for up to as many as 100 BSs. More specifically, the MSC provides the following functions:
- Mobility management for the subscribers (to register subscribers, to authenticate and authorize the subscribers for services and access to the network, to maintain the information on the temporary location of the subscribers so they can receive and originate voice calls).

In GSM, some of the functionality of the MSC is distributed to the Base Station Controller (BSC). In TDMA, the BSC and the MSC are integrated.

- Call setup services (call routing based on the called number). These calls can be to another mobile subscriber through another MSC, or to a landline user through the PSTN.
- Connection control services, which determine how calls are routed and establishes trunks to carry the bearer traffic to another MSC or to the PSTN.
- Service logic functions, which route the call to the requested service for the subscriber, such as an800 service, call forwarding, or voicemail.
- Transcoding functions, which decompress the voice traffic from the mobile device going to the PSTN and compresses the traffic going from the PSTN to the mobile device.

(3) Base Station (BS)

The BS is the component of the mobile wireless network access infrastructure that terminates the air interface over which the subscriber traffic is transmitted to and from a mobile station (MS). In GSM-based networks, the BS is called a base transceiver station (BTS).

(4) Radio Access Network (RAN)

The RAN identifies the portion of the wireless network that handles the radio frequencies (RF), radio resource management (RRM), which involves signaling, and the data synchronization aspects of transmission over the air interface.

In GSM-based networks, the RAN typically consists of BTSs and base station controllers (BSCs). User sessions are connected from a mobile station to a BTS, which connects to a BSC. The combined functions of the BTS and BSC are referred to as the base station subsystem (BSS)[4].

(5) Home Location Register (HLR)

The HLR is a database that contains information about subscribers to a mobile network that is maintained by a particular service provider. In addition, for subscribers of a roaming partner, the HLR might contain the service profiles of visiting subscribers.

The MSC uses the subscriber information supplied by the HLR to authenticate and register the subscriber. The HLR stores "permanent" subscriber information (rather than temporary subscriber data, which a VLR manages), including the service profile, location information, and activity status of the mobile user.

(6) Visitor Location Register (VLR)

The VLR is a database that is maintained by an MSC, to store temporary information about subscribers who roam into the coverage area of that MSC.

The VLR, which is usually part of an MSC, communicates with the HLR of the roaming subscriber to request data, and to maintain information about the subscriber's current location in the network.

(7) Authentication Center (AUC)

The AUC provides handset authentication and encryption services for a service provider. In most wireless networks today, the AUC is collocated with the HLR, and is often implemented as part of the HLR complex.

3. Model for IP Integration into Mobile Wireless

The standards for the integration of IP data networking with the existing telecommunications infrastructure are rapidly developing and beginning to be realized in today's production networks.

Figure 2-1 shows a model for IP integration based upon the current industry direction and

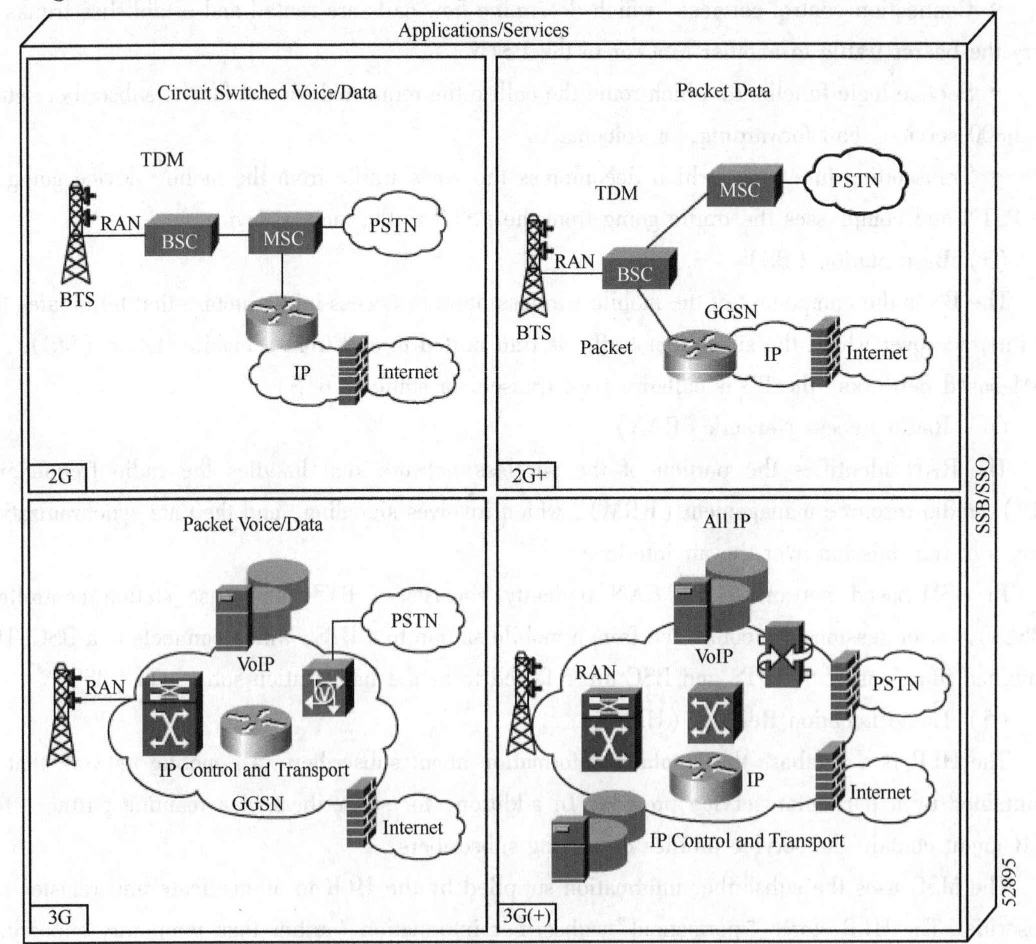

Figure 2-1 IP Integration Phases in Mobile Wireless

reflects some of the latest ideas within the Mobile Wireless Internet Forum (MWIF). The MWIF is a pre-standards consortium for service providers and suppliers to collaborate on the implementation of IP-based mobile wireless networks. The MWIF influences the standards bodies such as 3GPP and 3GPP2 to successfully adopt new implementations.

In particular, Figure 2-1 shows where Cisco Systems' GGSN product for GSM networks fits into the model.

The top two quadrants in Figure 2-1 show where we are today in the telecommunications and IP data services infrastructures. The first quadrant represents the first phase of these infrastructures based on circuit-switched voice and data services. The beginnings of a core IP transport for voice and data integration can be built using Cisco Systems V. 110 solutions.

The second quadrant depicts the implementation phase of 2G + technologies, such as GPRS, supporting higher transmission speeds. In this quadrant, the Cisco Systems GGSN provides IP packet data services. It acts as an IP gateway for access to the Internet and other public and private data networks for traffic that is initiated in a GSM-based mobile environment. The services anticipated in this phase include implementingalways-on data services and enabling operators to charge by packet rather than connect time. Similar services are supported by Packet Data Services Node (PDSN), for CDMA-based wireless networks.

The third quadrant represents phase three of the integration of IP networking where voice and data are consolidated onto a packet-based infrastructure from the RAN or radio network control (RNC) outward. This is considered a 3G solution. Phase three enables integrated voice and data applications and reduces costs. In addition, some of the components or functions of the MSC are distributed.

The fourth quadrant represents the final phase, which includes 3G services plus the implementation of IP-based radio and mobility components to develop a true end-to-end, all-IP wireless network solution.

New Words

convergence	[kən'vɜːdʒəns]	*n.* 趋同；集收敛；集合，会聚；［气］辐合
infrastructure	['ɪnfrəˌstrʌktʃə]	*n.* 基础结构，基础架构；基础设施；基础建设
differentiate	[ˌdɪfə'renʃɪeɪt]	*v.* 区分，辨别；使不同
ubiquitous	[juː'bɪkwɪtəs]	*adj.* 无所不在的，普遍存在的
assimilation	[əˌsɪmɪ'leɪʃən]	*n.* 吸收，消化；［生］同化作用
subscriber	[sʌbs'kraɪbə]	*n.* 用户，订户
seamless	['siːmlɪs]	*adj.* 无缝的；无漏洞的
roam	[rəʊm]	*v.* & *n.* 漫游，漫步
cellular	['seljʊlə]	*adj.* 蜂窝状的；细胞的；多孔的
analog	['ænəlɒg]	*adj.* ［电］模拟的；（钟表）有长短针的
implement	['ɪmplɪmənt]	*v.* 实施，执行；使生效，实现
hub	[hʌb]	*n.* （电器面板上的）电线插孔；［计］集线器

authenticate	[ɔːˈθentɪkeɪt]	*v.* 使生效；鉴别；证明是真实的
encryption	[ɪnˈkrɪpʃən]	*n.* 编密码；加密
decompress	[ˌdiːkəmˈpres]	*v.* 解压
compress	[kəmˈpres]	*v.* 压紧；压缩；精简
synchronization	[ˌsɪŋkrənaɪˈzeɪʃən]	*n.* 同步，使时间互相一致，同时性
profile	[ˈprəʊfaɪl]	*n.* 人物简介；外形，轮廓
consortium	[kənˈsɔːtjəm]	*n.* 组合，共同体；财团
quadrant	[ˈkwɒdrənt]	*n.* 象限仪，四分仪；四分之一圆
gateway	[ˈgeɪtweɪ]	*n.* 网关；入口；途径

Notes

[1] As suggested by its name, mobile wireless communication addresses those wireless technologies that support mobility of a subscriber, which provide seamless and real-time services without interruption.
顾名思义，移动无线通信用以称呼那些支持用户流动性的无线技术，它能够提供无缝连接和不间断的实时业务。

[2] AMPS 是第一代蜂窝技术，使用单独的频带，或者说“信道”，为每次对话服务。它因此需要相当的带宽来支持一个大数量的用户群体。在通用术语中，AMPS 常常被当作更早的“0G”改进型移动通信服务，只不过 AMPS 使用更多的计算功率来选择频谱、切换到 PSTN 线路的通话，以及处理登记和呼叫建立。真正将 AMPS 从更早的 0G 系统中区分出来的是最后的呼叫建立功能。在 AMPS 中，蜂窝中心可以根据信号强度灵活地分配信道给每个手持终端，允许相同的频率在完全不同的位置复用，并且不会有干扰。这使得在一个地区内，同时支持大数量的手持终端成为可能。AMPS 的创始者们发明了“蜂窝”这个术语正是因为它在一个系统里使用的都是小的六边形“蜂窝”形状。

[3] Authentication Center（AUC）：称为鉴权中心，是 GSM 系统中的安全管理单元，存储鉴权算法和密钥，保证各种保密参数的安全性，向 HLR（归属用户位置寄存器）提供鉴权参数。鉴权参数包括三组：RAND（Random Number，随机数），SRES（Sign Response，符号响应），Kc（Ciphering Key，加密密钥）。

[4] Base Station Subsystem（BSS）：是指基站子系统，是移动通信系统中与无线蜂窝网络关系最直接的基本组成部分。在整个移动网络中基站主要起中继作用。基站与基站之间采用无线信道连接，负责无线发送、接收和无线资源管理。而主基站与移动交换中心（MSC）之间常采用有线信道连接，实现移动用户之间或移动用户与固定用户之间的通信连接。

Questions for Discussion

1. What are the basic elements of wireless technologies?
2. What components is a typical cellular telecommunications network made up of?
3. What is MSC? What does it serve as in the mobile wireless network infrastructure?
4. What is HLR and VLR? What are their functions?

Text B

2G, 3G, 4G, 4G LTE, 5G—What Are They?

Quite simply, the "G" stands for Generation, as in the next generation of wireless technologies. Each generation is supposedly faster, more secure and more reliable. The reliability factor is the hardest obstacle to overcome. 1G was not used to identify wireless technology until 2G, or the second generation, was released. That was a major jump in the technology when the wireless networks went from analog to digital. It's all uphill from there. 3G came along and offered faster data transfer speeds, at least 200 Kilobits per second, for multi-media use and was a long time standard for wireless transmissions regardless of what you heard on all those commercials.

It is still a challenge to get a true 4G connection, which promises upwards of a 1Gbit/s, Gigabit per second, transfer rate if you are standing still and in the perfect spot. 4G LTE[1] comes very close to closing this gap. True 4G on a wide spread basis may not be available until the next generation arrives. 5G?

1. What Are the Standards of the G's?

Each of the Generations has standards that must be met to officially use the G terminology. Those standards are set by, you know, those people that set standards. The standards themselves are quite confusing but the advertisers sure know how to manipulate them. I will try to simplify the terms a bit.

1G. A term never widely used until 2G was available. This was the first generation of cell phone technology. Simple phone calls were all it was able to do.

2G. The second generation of cell phone transmission. A few more features were added to the menu such as simple text messaging.

3G. This generation set the standards for most of the wireless technology we have come to know and love. Web browsing, e-mail, video downloading, picture sharing and other smartphone technology were introduced in the third generation. 3G should be capable of handling around 2 Megabits per second.

4G. The speed and standards of this technology of wireless needs to be at least 100 Megabits per second and up to 1 Gigabit per second to pass as 4G. It also needs to share the network resources to support more simultaneous connections on the cell. As it develops, 4G could surpass the speed of the average wireless broad and home Internet connection. Few devices were capable of the full throttle when the technology was first released. Coverage of true 4G was limited to large metropolitan areas. Outside of the covered areas, 4G phones regressed to the 3G standards. When 4G first became available, it was simply a little faster than 3G. 4G is not the same as 4G LTE which is very close to meeting the criteria of the standards. The major wireless networks were not actually lying to anyone when 4G first rolled out; they simply stretched the truth a bit. A 4G phone had to comply with the standards but finding the network resources to fulfill the true standard was difficult. You were buying

4G capable devices before the networks were capable of delivering true 4G to the device. Your brain knows that 4G is faster than 3G so you pay the price for the extra speed. The same will probably be true when 5G hits the markets.

4G LTE. Long Term Evolution—LTE sounds better. This buzzword is a version of 4G that is fast becoming the latest advertised technology and is getting very close to the speeds needed as the standards are set. When you start hearing about LTE Advanced, then we will be talking about true fourth generation wireless technologies because they are the only two formats realized by the International Telecommunications Union as True 4G at this time. But forget about that because 5G is coming soon to a phone near you. Then there is XLTE[2] which is a bandwidth charger with a minimum of double the bandwidth of 4G LTE and is available anywhere the AWS spectrum is initiated. Verizon, T-Mobile and Sprint have all advanced to the LTE technology with each carrier adding their own combination of wireless technologies to enhance the spectrum.

5G. There are rumors of 5G being tested although the specifications of 5G have not been formally clarified. We can expect that new technology to be rolled out around 2020 but in this fast-paced world it will probably be much sooner than that. Seems like a long ways away but time flies and so will 5G at speeds of 1 – 10Gbit/s.

2. Understanding 4G Technology Standards

The term "4G" references to a new speed standard in Internet connectivity, but what exactly does the term mean? What is 4G?

Most users are familiar with "4G" standards, as most smart phones use this communications standard. 4G simply means "fourth generation" in reference to the evolution of data transfer technologies. The first generation of mobile technology (1G) came in 1981 with analog transmission, and in 1992 was 2G appeared in the form of digital information exchange. 3G made its debut in 2001, and included multi-media support along with a peak transfer rate of at least 200 kilobits per second. True 4G support is here. It is no surprise, then, that 4G means "fourth generation" and represents a number of improvements over the 3G technology.

3. Who Sets the 4G Standards?

4G technology is meant to provide what is known as "ultra-broadband" access for mobile devices, and the International Telecommunications Union-Radio communications sector (ITU-R) created a set of standards that networks must meet in order to be considered 4G, known as the International Mobile Telecommunications Advanced (IMT-Advanced) specification.

4. What Are the 4G Standards?

First, 4G networks must be based on an all Internet protocol (IP) packet switching instead of circuit-switched technology, and use OFMDA multi-carrier transmission methods or other frequency-domain equalization (FDE) methods instead of current spread spectrum radio technology. In addition, peak data rates for 4G networks must be close to 100 megabits per second for a user on a highly mobile network and 1 gigabit per second for a user with local wireless access or a nomadic connection. True 4G must also be able to offer smooth handovers across differing networks without data loss and provide high quality of service for next-gen media.

One of the most important aspects of 4G technology is the elimination of parallel circuit-switched and packet-switched network nodes using Internet protocol version 6 (IPv6). The currently used standard, IPv4, has a finite limitation on the number of IP addresses that can be assigned to devices, meaning duplicate addresses must be created and reused using network address translation (NAT), a solution that only masks the problem instead of definitively solving it. IPv6 provides a much larger number of available addresses, and will be instrumental in providing a streamlined experience for users.

5. Understanding LTE Technology Standards

The market for wireless data transfer is quickly evolving as global rather than local solutions are pursued—solutions offering robust data transfer rates, predictable spikes in traffic and are interoperable. One of the fastest-growing standards in the competition for ownership of the new fourth generation or "4G" market is LTE technology, which promises a number of improvements over current mobile and data terminal service.

LTE actually stands for "long term evolution", and its full name is 3GPP LTE, with the 3GPP standing for the 3rd Generation Partnership Project, which has been developing the technology's release documents. Often, LTE is marketed as 4G technology by companies that package it as part of their wireless or mobile service, but the standard is better thought of as "3.9G" as it does not yet meet the requirements set out by the ITU-R for 4G, which includes minimum upload and download rates for networks and defines how connections must be established. A new version of LTE technology, LTE Advanced, does satisfy the requirements of a true 4G network and is expected within the next year.

So, what exactly is LTE? It's based on GSM[3]/EDGE and UMTS/HSPA network technologies, and provides an increase to both capacity and speed using new techniques for modulation. It provides peak download rates of 300 megabits per second, upload rates of 75 megabits per second and a transfer latency of less than five milliseconds. It can also manage multi-cast and broadcast streams and handle quick-moving mobile phones. Its Evolved Packet Core (EPC), IP-based network architecture, allows for seamless handovers for voice and data to older model cell towers that use GSM, UMTS or CDMA2000 technology. In addition, it can scale from 1.4 MHz to 20 MHz carrier bandwidths and supports both time-division and frequency division duplexing. Overall, the new architecture of LTE technology means lower operating costs along with greater overall data and voice capacity.

6. Understanding 5G Technology Standards

5G stands for the fifth generation of wireless technologies and it will be faster than 4G. That is a no-brainer but how much faster is the question. The details are a bit sketchy at this point but the speeds are supposed to be upwards of 1 to 10Gbit/s compare to the 4G standards which are

100Mbit/s up to 1Gbit/s. Yeah a lot faster. But will those speeds ever be realized is another question that we will find out sometime around the year 2020. That is the expected date of the rollout. I am fairly confident that the term 5G will be used long before the official release. You will be hearing terms like— "our new 5G network" and "the newest 5G phone is now available" long before true 5G speeds are actually achieved if the 4G rollout is an example.

How fast will 5G be? A picture is worth a thousand words and this one from GSMA Intelligence says a lot. 5G is going to be fast (see Figure 2-2). Fast enough for you? It is supposed to be fast enough for everyone everything and the IoT (The Internet of Things). 5G usage goes way beyond your smart phone and devices. This will be what drives your cars; it will allow machines to communicate and pretty much anything else that will benefit from being connected.

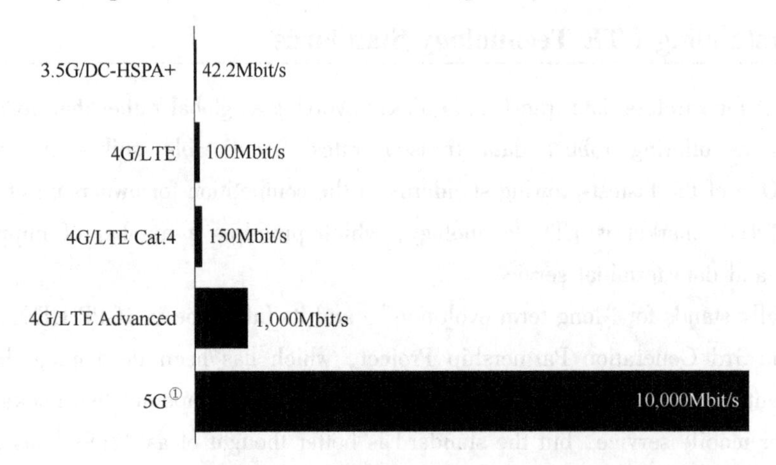

Figure 2-2 Maximum Theoretical Downlink Speed by Technology Generation (Mbit/s)

①10 Gbit/s is the minimum theoretical upper limit speed specified for 5G.

New Words

wireless	['waɪələs]	*adj.* 无线的；无线电的
manipulate	[mə'nɪpjʊleɪt]	*v.* 操纵；操作；巧妙地处理；篡改
simultaneous	[ˌsɪm(ə)l'teɪnɪəs]	*adj.* 同时的；联立的；同时发生的
surpass	[sə'pɑːs]	*v.* 超越；胜过，优于；非……所能办到或理解
throttle	['θrɒt(ə)l]	*n.* 节流阀
metropolitan	[metrə'pɒlɪt(ə)n]	*n.* 大城市人 *adj.* 大都市的
buzzword	['bʌzwɜːd]	*n.* 流行词
bandwidth	['bændwɪtθ]	*n.* [电子] [物] 带宽；[通信] 频带宽度
available	[ə'veɪləb(ə)l]	*adj.* 可获得的；可购得的；可找到的；有空的
megabit	['megəbɪt]	*n.* [计] 兆位；百万位
gigabit	['gɪgəbɪt; 'dʒ-]	*n.* [计] 千兆比特
streamline	['striːmlaɪn]	*n.* 流线；流线型 *v.* 把……做成流线型
spike	[spaɪk]	*n.* 长钉，道钉；钉鞋；细高跟 *v.* 阻止；以大钉钉牢

interoperable[ɪntər'ɒp(ə)rəb(ə)l] *adj.* 彼此协作的；能共同操作的；能共同使用的
rollout ['rəʊlaʊt] *n.* 首次展示；[航] 滑跑
downlink ['dəʊnˌlɪŋk] *n.* 下行线；向下链路；向地传输

Notes

[1] LTE（Long Term Evolution）：长期演进是由 3GPP（The 3rd Generation Partnership Project，第三代合作伙伴计划）组织制定的 UMTS（Universal Mobile Telecommunications System，通用移动通信系统）技术标准的长期演进，于 2004 年 12 月在 3GPP 多伦多会议上正式立项并启动。

[2] XLTE：2014 年 05 月 19 日美国运营商 Verizon 正式推出了更快的 4G LTE 网络品牌，命名为"XLTE"，该网络将比现有 LTE 网络快一倍。据 Verizon 官网介绍，XLTE 被称作全美规模最大、最可靠的 LTE 网络，能够提供两倍于现有宽带网速的体验，覆盖超过 97% 的美国境内用户。

[3] GSM 是全球移动通信系统（Global System for Mobile Communications）的简称。它的空中接口采用时分多址技术。自 20 世纪 90 年代中期投入商用以来，被全球超过 100 个国家采用。GSM 标准的设备占据当前全球蜂窝移动通信设备市场 80% 以上。GSM 是当前应用最为广泛的移动电话标准。全球超过 200 个国家和地区超过 10 亿人正在使用 GSM 电话。

Questions for Discussion

1. What is 3G?
2. What exactly does LTE mean?
3. What does 5G stand for?

Unit 3

Text A

Internet & Communication

1. What Is Internet?

The Internet has revolutionized the way we work and play. It allows us to communicate, to share data and to seek information in a matter of seconds. All this is possible through the use of computers and networks.

The Internet is a global network of computers. All computer devices (including PCs, laptops, game consoles and smartphones) that are connected to the Internet form part of this network. Added together, there are billions of computers connected to the Internet, all able to communicate with each other. Today, the Internet is a massive part of our daily lives (see Figure 3-1).

Figure 3-1

2. How Did the Internet Originate?

In the 1950s, the United States Defence Department formed several agencies, such as the Advanced Research Projects Agency (ARPA, now known as DARPA) with the purpose of developing technology. However, since they were based at universities around the country, ARPA's scientists could not easily communicate or share information. To solve this problem, ARPA created a network of computers, which they called ARPANET. Realizing how useful ARPANET was, other organizations built their own networks. However, these individual networks could not easily communicate with each other.

In the 1970s, a protocol was developed. Called TCP/IP, this protocol allowed the separate networks to communicate with each other. The joining of these individual networks created a huge wide area network (WAN) which came to be known as the Internet.

Since then the use of the internet by organizations and individuals has grown year upon year. In the beginning, ARPANET consisted of just four computers. Now billions of computers are connected to the internet. When we connect to the internet, we are said to be 'online'. Today the internet has many online facilities, for example:

- Communication via e-mail and VoIP;
- Sharing of information such as text, images, sounds and videos;
- Storage of information;
- Streaming television programmes, films, videos, sounds and music;
- Playing online games;
- Shopping;
- Social networking;
- Banking

Most of these online facilities are available through the use of websites on the World Wide Web.

3. What Is World Wide Web?

The Internet is a global network of computers. The World Wide Web is the part of the Internet that can be accessed through websites. Websites consist of webpages which allow you to see information.

Websites are accessed using a web browser. A browser is a program designed to display the information held on a website. Every website has an address at which it can be found, a bit like a house address.

A website's address is known as its URL. A website can be visited by typing its URL into a web browser. Each address contains the prefix "http:" which tells the computer to use the hypertext transfer protocol for communicating with the website. [1] The browser then connects to the internet, finds the website at its address and downloads the information stored there onto our computer for us to view. Websites and webpages are joined together using hyperlinks. Clicking on a hyperlink takes us to another site or page.

4. Transferring Information via the Internet

The Internet is a global network of computers, some of which are called web servers. A web server is a computer which holds websites for other computers linked to the Internet to access. Holding a website is known as "hosting". A web server may host one or many websites and webpages. Sending information to a web server is known as uploading. Receiving information from a web server is known as downloading.

When you make a telephone call, a direct connection is formed between you and the person you are calling. While you are making the call, no one else can communicate with you. A web server needs to be able to communicate with many different computers at the same time. When information is uploaded to, or downloaded from, a web server it is broken up into tiny pieces called data packets. Each packet is a very short communication between the client computer and the web server. Because each communication lasts only milliseconds, the web server can seemingly communicate with many computers at the same time. It is a bit like having several conversations at the same time, but only saying one word to each person in turn.

5. Using HTML to Create Website

All web pages on the Internet are created using a language called Hypertext Markup Language (HTML)[2]. HTML describes:

- What information appears on a webpage.
- How it appears on the page (formatting).
- Any links to other pages or sites.

HTML can be written in specialist software, or in a simple text editor like Notepad. As long as the document is saved with the file extension ".html" it can be opened and viewed as a webpage from a browser. This example HTML code displays a message on a webpage (see Figure 3-2):

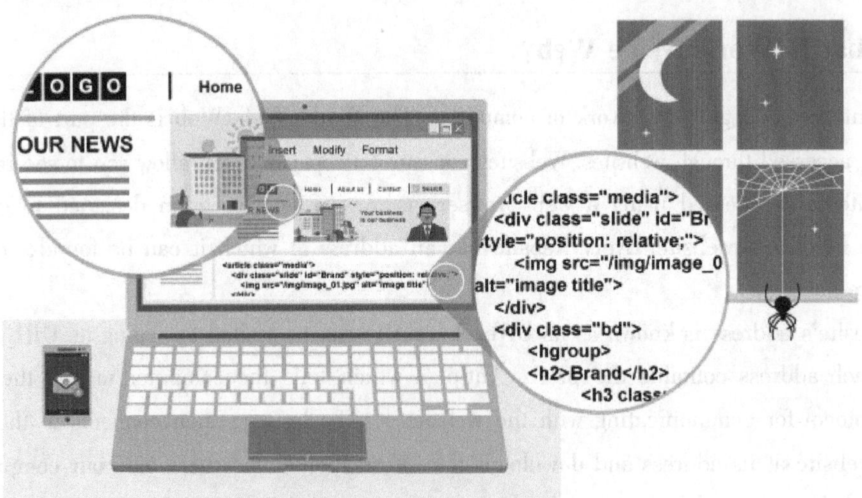

Figure 3-2

```
<html>
    <body>
      <h1>Hello world</h1>
      <p>This is my first webpage</p>
    </body>
</html>
```

The code uses tags to describe the appearance of the information:

- <html> states that the document is an HTML document;
- <body> states that the information appears in the body of the page;
- <h1> states that the following text appears as a prominent heading;
- <p> states that this is the beginning of a new paragraph.

6. Using E-mail to Communicate

Two of the main ways in which people use the Internet for communication are e-mail and VoIP[3].

E-mail (short for electronic mail) is the digital equivalent of sending a letter. Each e-mail has a sender, a receiver and a message. The big difference is that instead of waiting for our posted letter to be delivered by the post office, an e-mail is sent and received almost immediately (see Figure 3-3).

Figure 3-3

Advantages and Disadvantages of e-mail

Aside from speed, e-mail has several other advantages:

- It costs virtually nothing to send an e-mail, whereas you need to buy stamps to post a letter.
- The same e-mail can be sent to many people at the same time.
- Assets, such as text, videos or sound clips, can be attached to e-mail.
- A record is kept of each sent e-mail, so it is possible to refer back to check what was sent.
- When e-mails are sent, the recipient doesn't have to be there. E-mails can be sent late at night and the recipient will see it when they check their e-mail the following day. This has great benefits when sending e-mails to another part of the world.
- E-mails can be sent at any time or day of the year. Post is not usually delivered on Sundays or bank holidays.
- E-mails can be sent and received on various Internet connected devices, eg PCs, laptops, games consoles, tablets or smartphones.
- E-mails can be automatically forwarded on to another address.

Using e-mail has its disadvantages too:

- The recipient can only receive the e-mail if they are connected to the Internet.
- E-mails can sometimes contain viruses in the form of attachments.
- Spam e-mails can be a problem. So can phishing e-mails, which are designed to trick people into giving away personal information.
- Because e-mails can be delivered to Internet-connected digital devices anywhere, they can be hard to get away from.

7. Using VoIP and Video Conferencing

A video conference is live video streamed over the Internet so that people can communicate face to face without being in the same room. Although this was mainly used by businesses and academics at first, video conferencing is now used socially by many people. Voice over Internet Protocol (VoIP) is the technology that allows us to video conference. Many companies offer free VoIP services, including Skype, Apple FaceTime and Google Hangouts.

To video conference with another person, both people need an Internet-connected computer with a monitor, webcam, microphone and speakers. The webcam allows video images to be sent, which are seen on the monitor, and the microphone allows sound to be transmitted, which is heard through speakers. A program called a VoIP client handles the communication.

Video conferencing technology is built into many laptops, tablets and smartphones, and VoIP apps are available to use, often for free.

Advantages and Disadvantages of Video Conferencing

Video conferencing has several advantages:

- Seeing as well as hearing the other person.
- Showing others what is going on around us.
- Reducing time to travel to see and speak with someone. This has even greater benefits if the

other person is on the other side of the world.

- Saving money, in travel costs.
- The ability to video conference several people in different locations, at the same time.

Using video conferencing also has its disadvantages:

- Everyone who wants to video conference must have suitable hardware and software.
- Video conferencing from smartphones can be expensive because of the high data usage.
- Streaming video two ways requires a good deal of bandwidth. You might struggle to video conference if your connection to the Internet is of poor quality.

New Words

laptop	[ˈlæptɒp]	*n.* 便携式计算机
console	[kənˈsəʊl]	*n.* 控制台，操纵台；演奏台
smartphone	[smɑːtfəʊn]	*n.* 智能手机
massive	[ˈmæsɪv]	*adj.* 大量的，重的，大块的
hyperlink	[ˈhaɪpəlɪŋk]	*n.* 超链接
upload	[ˈʌpˌləʊd]	*v.* 上传
millisecond	[ˈmɪlɪsekənd]	*n.* 毫秒
extension	[ɪkˈstenʃən]	*n.* 伸展，扩大；电话分机
virtually	[ˈvɜːtʃʊəli]	*adv.* 几乎；实际上
phish	[fɪʃ]	*v.* 网络钓鱼
monitor	[ˈmɒnɪtə]	*n.* 监视器；显示屏
recipient	[rɪˈsɪpɪənt]	*n.* 接受者；容器，容纳者
webcam	[ˈwebkæm]	*n.* 网络摄像头
client	[ˈklaɪənt]	*n.* ［经］客户；顾客；委托人

Notes

[1] Each address contains the prefix "http:" which tells the computer to use the hypertext transfer protocol for communicating with the website.

hypertext transfer protocol 是互联网上应用最为广泛的一种网络协议。所有的 WWW 文件都必须遵守这个标准。设计 HTTP 最初的目的是提供一种发布和接收 HTML 页面的方法。1960 年美国人 Ted Nelson 构思了一种通过计算机处理文本信息的方法，并称之为超文本（hypertext），这成了 HTTP 超文本传输协议标准架构的发展根基。Ted Nelson 组织协调万维网协会（World Wide Web Consortium）和互联网工程工作小组（Internet Engineering Task Force）共同合作研究，最终发布了一系列的 RFC，其中著名的 RFC 2616 定义了 HTTP 1.1。

[2] Hypertext Markup Language（HTML）即超文本标记语言。它是为网页创建和其他可在网页浏览器中看到的信息设计的一种标记语言。Extensible Hypertext Markup Language（XHTML），即可扩展超文本标记语言，是一种新兴的网页设计和制作语言。XHTML 是在 HTML 基础上发展起来的，同时吸取了可扩展标记语言（Extensible Markup

Language，XML）的语法严谨的优点。因此，XHTML 比 HTML 具有更加严谨的语法，能够为众多品牌的 Web 浏览器研发提供规范的技术标准。XHTML 的可扩展性和灵活性将适应未来网络应用的更多需求。

[3] VoIP（Voice over Internet Protocol）简而言之就是将模拟信号（Voice）数字化，以数据封包（Data Packet）的形式在 IP 网络（IP Network）上做实时传递。VoIP 最大的优势是能广泛地采用 Internet 和全球 IP 互连的环境，提供比传统业务更多、更好的服务。VoIP 可以在 IP 网络上便宜地传送语音、传真、视频和数据等业务，如统一消息业务、虚拟电话、虚拟语音/传真邮箱、查号业务、Internet 呼叫中心、Internet 呼叫管理、电话视频会议、电子商务、传真存储转发和各种信息的存储转发等。

Questions for Discussion

1. What kind of online conveniences can we enjoy through use of websites on the World Wide Web?
2. Can you list some disadvantages of e-mails? What are they?
3. What are major advantages of video conferencing?
4. In what aspects do you think Internet will further influence us?

Text B

Types of Internet Communications

If you use the Internet, then you probably use Internet-based communications to contact family, friends or co-workers. From sending an instant message to a friend, to e-mailing co-workers, to placing phone calls, to conducting video conferences, the Internet offers a number of ways to communicate.

The advantages of Internet-based communications are many. Since you're already paying for an Internet account (or your employer is), you can save money on phone calls by sending someone an instant message or by using VoIP instead of standard local telephone services. Of course, no technology is without a downside and Internet-based communications has plenty, such as viruses, privacy issues and spam.

Like all technologies (and especially technology tied to the Internet), the way we can communicate online is constantly evolving. In this week's *Did You Know...?* article we'll take a look at some of the most popular forms of Internet-based communications.

1. Instant Messaging

One of the fastest-growing forms of Internet communications is instant messaging[1], or IM. Think of IM as a text-based computer conference between two or more people. An IM communications service enables you to create a kind of private chat room with another individual in order to communicate in real-time over the Internet. Typically, the IM system alerts you whenever somebody on your buddy or contact list is online. You can then initiate a chat session with that particular individual.

One reason that IM has become so popular is its real-time nature. Unlike e-mail, where you will wait for the recipient to check his or her e-mail and send a reply, if a person you want to reach is online and available in your IM contact list, your message appears instantly in a window on their screen.

While IM is used by millions of Internet users to contact family and friends, it's also growing in popularity in the business world. Employees of a company can have instant access to managers and co-workers in different offices and can eliminate the need to place phone calls when information is required immediately. Overall, IM can save time for employees and help decrease the amount of money a business spends on communications.

Some problems and issues associated with IM include spim and virus propagation. Spim is the IM equivalent of spam and is perpetuated by bots that harvest IM screen names off of the Internet and simulate a human user by sending spim to the screen names via an instant message. The spim typically contains a link to a website that the spimmer is trying to market. Spim is a bit more intrusive than spam due to the nature of IM itself. These advertisements and junk messages will pop-up in your IM window and you need to deal with the messages immediately, where with e-mail you can usually filter a lot of it out and deal with it later. Additionally, viruses and Trojans can be spread through IM channels. These malicious programs are usually spread when an IM user receives a message that links to a website where the malicious code is downloaded. The message will appear to be from a known IM contact, which is why recipients are more likely to click the hyperlink and download the file. Using safe chat rules (such as never clicking the link) and keeping an updated anti-virus program on your system will help reduce the chances of becoming infected by malicious programs being spread through instant messaging.

2. Internet Telephone & VoIP

Internet telephony consists of a combination of hardware and software that enables you to use the Internet as the transmission medium for telephone calls. For users who have free, or fixed-price Internet access, Internet telephony software essentially provides free telephone calls anywhere in the world. In its simplest form, PC-to-PC Internet telephony can be as easy as hooking up a microphone to your computer and sending your voice through a cable modem to a person who has Internet telephony software that is compatible with yours. This basic form of Internet telephony is not without its problems, however. Connecting this way is slower than using a traditional telephone, and the quality of the voice transmissions is also not near the quality you would get when placing a regular phone call.

VoIP is another Internet-based communications method which is growing in popularity. VoIP hardware and software work together to use the Internet to transmit telephone calls by sending voice data in packets using IP rather than by traditional circuit transmissions, called PSTN (Public Switched Telephone Network). The voice traffic is converted into data packets then routed over the Internet, or any IP network, just as normal data packets would be transmitted. When the data packets reach their destination, they are converted back to voice data again for the recipient.

Much like finding an Internet service provider (ISP) for your Internet connection, you will need to use a VoIP provider. Some service providers may offer plans that include free calls to other subscribers on their network and charge flat rates for other VoIP calls based on a fixed number of calling minutes. You most likely will pay additional fees when you call long distance using VoIP. While this sounds a lot like regular telephone service, it is less expensive than traditional voice communications, starting with the fact that you will no longer need to pay for extras on your monthly phone bill.

3. E-mail

Short for electronic mail, e-mail is the transmission of messages over communications networks. The messages can be notes entered from the keyboard or electronic files stored on disk. Most mainframes, minicomputers and computer networks have an e-mail system. Some e-mail systems are confined to a single computer system or network, but others have gateways to other computer systems, enabling you to send electronic mail anywhere in the world.

Using an e-mail client (software such as Microsoft Outlook or Eudora), you can compose an e-mail message and send it to another person anywhere, as long as you know the recipient e-mail address. All online services and ISPs offer e-mail, and support gateways so that you can exchange e-mail with users of other systems. Usually, it takes only a few seconds for an e-mail to arrive at its destination. This is a particularly effective way to communicate with a group because you can broadcast a message or document to everyone in the group at once.

One of the biggest black clouds hanging over e-mail is spam. Though definitions vary, spam can be considered any electronic junk mail (generally e-mail advertising for some product) that is sent out to thousands, if not millions, of people. Often spam perpetrates the spread of e-mail Trojans and viruses. For this reason, it's important to use an updated anti-virus program, which will scan your incoming and outgoing e-mail for viruses.

4. IRC

Short for Internet relay chat, IRC is a multi-user chat system that allows to people gather on "channels" or "rooms" to talk in groups or privately. IRC is based on a client/server model. That is, to join an IRC discussion, you need an IRC client (such a mIRC) and Internet access. The IRC client is a program that runs on your computer and sends and receives messages to and from an IRC server. The IRC server, in turn, is responsible for making sure that all messages are broadcast to everyone participating in a discussion. There can be many discussions going on at once and each one is assigned a unique channel. Once you have joined an IRC chatroom (chatroom discussions are designated by topics), you can type your messages in the public chatroom where all participants will see it, or you can send a private message to a single participant. With many IRC clients you can easily create your own chatroom and invite others to join your channel. You can also password protect your chatroom to allow for a more private discussion with just people whom you invite.

5. Video-Conferencing

Video-conferencing is a conference between two or more participants at different sites by using computer networks to transmit audio and video data. Each participant has a video camera, microphone and speakers connected on his or her computer. As the two participants speak to one another, their voices are carried over the network and delivered to the other's speakers, and whatever images appear in front of the video camera appear in a window on the other participant's monitor.

In order for video-conferencing to work, the conference participants must use the same client or compatible software. Many freeware and shareware video-conferencing tools are available online for download, and most Web cameras also come bundled with video-conferencing software. Many newer video-conferencing packages can also be integrated with public IM clients for multipoint conferencing and collaboration. In recent years, video-conferencing has become a popular form of distance communication in classrooms, allowing for a cost efficient way to provide distance learning, guest speakers, and multi-school collaboration projects. Many feel that video-conferencing provides a visual connection and interaction that cannot be achieved with standard IM or e-mail communications.

6. SMS & wireless communications

Short message service (SMS) is a global wireless service that enables the transmission of alphanumeric messages between mobile subscribers and external systems such as e-mail, paging and voice-mail systems. Messages can be no longer than 160 alpha-numeric characters and must contain no images or graphics. Once a message is sent, it is received by a Short Message Service Center (SMSC), which must then get it to the appropriate mobile device or system. As wireless services evolved, multimedia messaging service (MMS) was introduced and provided a way to send messages comprising a combination of text, sounds, images and video to MMS capable handsets.

Communication on wireless devices such as mobile phones and PDAs is frequently changing. Today you can use your wireless device to not only make phone calls, but to send and receive e-mail and IM. While you can use e-mail, IRC or IM for free if you have an Internet account, you will end up paying fees to your mobile carrier to use these services on a wireless device.

New Words

virus	[ˈvaɪrəs]	*n.* 病毒；恶毒；毒害
spam	[spæm]	*v.* 刷屏，垃圾邮件
evolve	[ɪˈvɒlv]	*v.* 发展，进展；进化；逐步形成
initiate	[ɪˈnɪʃɪeɪt]	*v.* 开始，创始；发起；使初步了解
propagation	[ˌprɒpəˈgeɪʃən]	*n.* 传播；繁殖；增殖
perpetuate	[pəˈpetʃʊeɪt]	*v.* 使不朽；保持 *adj.* 长存的
intrusive	[ɪnˈtruːsɪv]	*adj.* 侵入的；打扰的
malicious	[məˈlɪʃəs]	*adj.* 恶意的；恶毒的；蓄意的；怀恨的
mainframe	[ˈmeɪnfreɪm]	*n.* [计] 主机；大型机

videoconference	[ˌvɪdɪəʊˈkɒnfərəns]	*n.* 可视会议
compatible	[kəmˈpætɪb(ə)l]	*adj.* 兼容的；能共处的；可并立的
freeware	[ˈfriːweə]	*n.* 免费软件
shareware	[ˈʃeəweə]	*n.* 共享软件
collaboration	[kəlæbəˈreɪʃn]	*n.* 合作；勾结；通敌
alphanumeric	[ˌælfənjuːˈmerɪk]	*adj.* [计] 字母数字的

Notes

[1] instant messaging（IM）是一种可以让使用者在网络上建立某种私人聊天室（chatroom）的实时通信服务。目前，在互联网上受欢迎的即时通信软件包括腾讯 QQ、微信等。

Questions for Discussion

1. How many forms of Internet-based communications are introduced in the essay? What are they?
2. What are the problems and issues associated with IM?
3. What are the advantages of video-conferencing?

Unit 4

Text A

Five Reasons Why Fiber Is the Way of the Future

With all of the talk about fiber floating around the Internet, especially in relationship to Google's gigabit broadband service, many might be wondering just why fiber optics are so important. That is certainly a fair enough question to ponder, and the truth is that fiber optics have many advantages to offer customers and network architects alike. Before covering these individual benefits, it might be worth taking a step back and looking at the fundamental differences between traditional metal wires and fiber optics.

1. Metal Wires

Over one hundred and fifty years of data transmission metal wires have been sending electrical signals almost as long as the United States has been an independent nation, and that is saying something. The wires of today are far more sophisticated than those of yesteryear, but they still work in the same way: Electricity is applied to one end of the wires and it travels to the far end where it is received as a signal. Along the way, the signal strength degrades as the energy experiences a type of electrical friction called impedance. Impedance results in the signal decaying over distance and the wire becoming warmer, which can cause some problems that will be outlined later.

2. Fiber Optics

The power of light is the future. Fiber optics use specialized fibers that are capable of carrying transmissions made of pure light. Just like electrical wiring, a data transmission starts at one end of a fiber optic cable and transfers all the way to the end where it is received and decoded.

Benefit 1: Less Signal Degradation. While light does degrade over distance, the fact that the sun's light shines brightly upon us says one thing: Light travels a lot further than the constant EMP waves the sun emits because light degrades much slower than electricity does. In terms of a broadband service provider and their customers, this means that more residences and businesses can be served. Anyone who remembers the dawn of DSL will probably recall the limitations that were caused by the distances involved; if one did not live practically next door to a DSL[1] network node, they were out of luck. Things got better over time, but fiber optics has this problem solved from day one.

Benefit 2: More Untapped Overhead. Metal wires are already nearing their physical potential due to the heat caused by data transmissions. Too much load on any set of wires will result in those wires melting into a fine slag and becoming useless, and that is exactly what happens when too much is demanded of old wires. Fiber optics carry light, not electricity, and light has a negligible heat footprint in most cases. This is especially true of light created for fiber optic systems, which is far less potent than the UV light emitted by the sun.

Benefit 3: Easier Upgrades. While DSL and cable service providers have been able to systematically increase the performance of their networks over time, the entire act has been expensive. Often, whole segments of wires need to be dug up and replaced because of the heat problem. Fiber optics have the combination of substantially greater distance between network nodes and substations and more untapped overhead, which makes upgrades less of a hassle for network carriers. This in turn means less fees that have to be passed on to the consumers. A good case in point would be Verizon's FiOS[2] network, which has dramatically increased in terms of raw performance since its public debut in a manner that DSL and cable services have not.

Benefit 4: Fiber Is Green. Starting to think that sending data via electricity over metal wires is wasteful? If so, then you would be correct; data sent over metal wires takes dozens of times the energy that it takes to send a light signal. The additional substations and nodes needed to keep that signal strong over greater distances only adds to the woes of metal wires, and makes fiber optics look that much better by comparison. Furthermore, upgrades to networks are a lot less wasteful on the fiber optic side of the fence. What do telecoms and cable providers do with all of those "old" network nodes when they upgrade? Who knows, but fiber optics have fewer substations and nodes to replace and upgrade. Add to this the fact that the cables virtually never go bad, and it is simple to see why fiber optics are considered a green alternative to metal wires and electricity in any form.

Benefit 5: Psst... a Secret. Big DSL and Cable service providers already know how useful fiber optics are, and chances are good that their networks use fiber optics to get much closer to the homes and businesses of their consumers than they would like to let everyone know. In many cases, the fiber optic networks of DSL and/or cable providers actually go within a mile or so of many of the customers they serve. Why? Because they know that fiber optics are the most cost effective solution, and they know that by putting fiber close to the homes and businesses that they serve that they stand a very good chance of making the transition to an all-fiber network that much easier.

3. New form of Light Discovered, May Change the Future of Fiber Optics

Physicists in Ireland have turned quantum mechanics on its head. They have discovered a new form of light that challenges the very nature of how we have viewed and studied light up until now. It also spells big things for the future of fiber optics and data communication.

Until now, light has been seen as a fixed constant. Literally, your ability to see light is based on Planck's constant and angular momentum, a number that is a multiple based on Planck's mathematical equation which measures a beam of light. Now, this team at Trinity have found an angular momentum where a beam of light takes only half the value of Planck's constant. That qualifies it as a new form of light.

Professor John Donegan is interested to find out how we can use this new form of light in every day of life. One of the most immediate and impactful ways will be in fiber optics communication. Fiber optics is the foundation for computer data. By transmitting flashes of light through glass or plastic threads, it is now possible to send up to 400Gbit/s of data through a single channel. With this new discovery, that data transmission rate could go much higher, and be more secure.

Now, this idea that angular momentum wasn't fixed has been floating around for some time. This team finally was able to create a test environment to see if this could be achieved. To do so, they went back to the basics of the study of light and took a page from physicist Humphrey Lloyd and mathematician William Rowan Hamilton. In the 1830s, they observed conical refraction. Utilizing crystals, they saw how a ray of light could be formed into a single cone, or beam. Planck devised the math behind it, as it related photos, and now Prof. John Donegan, Assistant Professor Paul Eastham, and their team have taken it to the next level.

The discovery is still in its infancy, but already the Director of CRANN, Professor Stefano Sanvito knows the importance of this find, "This discovery is a breakthrough for the world of physics and science alike." It's just a matter of time before data communication companies come knocking on the doors of Trinity College Dublin[3], wanting to bring this new tech to the forefront of mass communications.

By *Meredith Placko*

New Words

fiber	[ˈfaɪbə]	*n.* 光纤
optics	[ˈɒptɪks]	*n.* 光学
sophisticated	[səˈfɪstɪkeɪtɪd]	*adj.* 复杂的；精致的；富有经验的
degrade	[dɪˈgreɪd]	*v.* 降低，贬低；使降级
friction	[ˈfrɪkʃən]	*n.* 摩擦；冲突，不和；摩擦力
impedance	[ɪmˈpiːdəns]	*n.* 阻抗，全电阻，电阻抗
decay	[dɪˈkeɪ]	*v.* 衰退，衰败，衰落
near	[nɪə]	*v.* 接近，临近
slag	[slæg]	*n.* 矿渣；熔渣
potent	[ˈpəutənt]	*adj.* 有效的，强有力的；烈性的
emit	[ɪˈmɪt]	*v.* 发出；发射；颁布
substantial	[səbˈstænʃəl]	*adj.* 大量的；牢固的；重大的
substation	[ˈsʌbsteɪʃən]	*n.* 变电站，变电所

untapped	[ʌnˈtæpt]	*adj.* 未开发的，未利用的
hassle	[ˈhæsəl]	*n.* 困难的事情；麻烦的事情；争论
node	[nəʊd]	*n.* 结点；（计算机网络的）节点
spell	[spel]	*v.* 导致；拼写；意味着
angular	[ˈæŋgjʊlə]	*adj.* 有角的；用角测量的，用弧度测量的
momentum	[məʊˈmentəm]	*n.* ［物］动量；势头；动力；要素
impactful	[ɪmˈpæktfʊl]	*adj.* 有效的，有力的
conical	[ˈkɒnɪkəl]	*adj.* 圆锥（形）的
refraction	[rɪˈfrækʃən]	*n.* 折射（程度）；折射角
cone	[kəʊn]	*n.* 圆锥体，锥形物；（松树的）球果

Notes

[1] DSL的中文名是数字用户线路，是以电话线为传输介质的传输技术组合。DSL技术在传递公用电话网络的用户环路上支持对称和非对称传输模式，解决了经常发生在网络服务供应商和最终用户间的"最后一公里"的传输瓶颈问题。

[2] Verizon公司是由美国大西洋贝尔和Nynex合并建立BellAtlantic后，独立电话公司GTE合并而成的，公司正式合并后，Verizon一举成为美国最大的本地电话公司、最大的无线通信公司，全世界最大的印刷黄页和在线黄页信息的提供商。FiOS（Fiber Optic Service）就是Verizon所提供的采用光纤电缆传输数据的数据通信服务。

[3] Trinity College Dublin"都柏林圣三一学院"，全称为：College of the Holy and Undivided Trinity of Queen Elizabeth near Dublin（伊丽莎白女王在都柏林附近神圣不可分割的三一学院）位于爱尔兰首都都柏林，是1592年英国女王伊丽莎白一世下令为"教化"爱尔兰而参照牛津大学和剑桥大学模式而兴建的。

Questions for Discussion

1. Why are metal wires inferior to fiber optics?
2. Why are fiber optics, compared with metal wires, much easier to upgrade?
3. Why do those offer fiber optic networks of DSL tend to have their fiber close to the homes of their customers as possible as they can?
4. What do you think of the discovery made by the team at Trinity?

Text B

Fiber to the x

Fiber to the x (FTTX[1]) is a generic term for any broadband network architecture using optical fiber to provide all or part of the local loop used for last mile telecommunications. As fiber optic cables are able to carry much more data than copper cables, especially over long distances, copper telephone networks built in the 20th century are being replaced by fiber.

FTTX is a generalization for several configurations of fiber deployment, arranged into two groups: FTTP/FTTH/FTTB (Fiber laid all the way to the premises/home/building) and FTTC/N (fiber laid to the cabinet/node, with copper wires completing the connection).

1. Definitions

The telecommunications industry differentiates between several distinct FTTX configurations. The terms in most widespread use today are:

- FTTP (fiber-to-the-premises): This term is used either as a blanket term for both FTTH and FTTB, or where the fiber network includes both homes and small businesses.
- FTTH (fiber-to-the-home): Fiber reaches the boundary of the living space, such as a box on the outside wall of a home. Passive optical networks and point-to-point Ethernet are architectures that deliver triple-play services over FTTH networks directly from an operator's central office.
- FTTB (fiber-to-the-building, -business, or -basement): Fiber reaches the boundary of the building, such as the basement in a multi-dwelling unit, with the final connection to the individual living space being made via alternative means, similar to the curb or pole technologies.
- FTTO (fiber-to-the-office): Fiber connection is installed from the main computer room/core switch to a special mini-switch (called FTTO Switch) located at the user's workstation or service points. This mini-switch provides Ethernet services to end user devices via standard twisted pair patch cords. The switches are located decentrally all over the building, but managed from one central point.
- FTTN/FTTLA (fiber-to-the-node, -neighborhood, or -last-amplifier): Fiber is terminated in a street cabinet, possibly miles away from the customer premises, with the final connections being copper. FTTN is often an interim step toward full FTTH (fiber-to-the-home) and is typically used to deliver "advanced" triple-play telecommunications services.
- FTTC/FTTK (fiber-to-the-curb/kerb, -closet, or -cabinet): This is very similar to FTTN, but the street cabinet or pole is closer to the user's premises, typically within 1, 000 feet (300 m), within range for high-bandwidth copper technologies such as wired Ethernet or IEEE[2] 1901 power line networking and wireless WiFi technology. FTTC is occasionally ambiguously called FTTP (fiber-to-the-pole), leading to confusion with the distinct fiber-to-the-premises system.

To promote consistency, especially when comparing FTTH penetration rates between countries, the three FTTH Councils of Europe, North America, and Asia-Pacific agreed upon definitions for FTTH and FTTB in 2006, with an update in 2009, 2011 and another in 2015. The FTTH Councils do not have formal definitions for FTTC and FTTN (see Figure 4-1).

Figure 4-1 illustrating how FTTX architectures vary with regard to the distance between the optical fiber and the end user. The building on the left is the central office; the building on the right is one of the buildings served by the central office. Dotted rectangles represent separate living or office spaces within the same building.

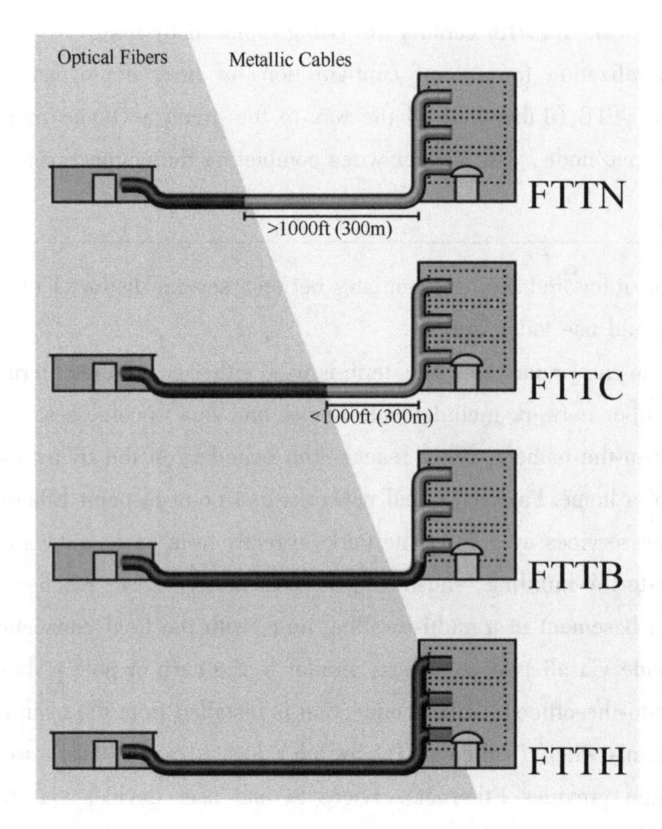

Figure 4-1

2. Benefits

While fiber optic cables can carry data at high speeds over long distances, copper cables used in traditional telephone lines and ADSL cannot. For example, the common form of gigabit Ethernet (1Gbit/s) runs over relatively economical category 5e, category 6 or augmented category 6 unshielded twisted-pair copper cabling but only to 100 m. However, 1 Gbit/s Ethernet over fiber can easily reach tens of kilometers. Therefore, FTTP has been selected by every major communications provider in the world to carry data over long 1 Gbit/s symmetrical connections directly to consumer homes. FTTP configurations that bring fiber directly into the building can offer the highest speeds since the remaining segments can use standard Ethernet or coaxial cable. Google Fiber provides speed of 1 Gbit/s.

Fiber is often said to be "future-proof" because the data rate of the connection is usually limited by the terminal equipment rather than the fiber, permitting substantial speed improvements by equipment upgrades before the fiber itself must be upgraded. Still, the type and length of employed fibers chosen, e. g. multimode vs. single-mode, are critical for applicability for future connections of over 1 Gbit/s.

FTTC (where fiber transitions to copper in a street cabinet) is generally too far from the users for standard Ethernet configurations over existing copper cabling. They generally use very-high-bit-rate digital subscriber line (VDSL) at downstream rates of 80 Mbit/s, but this falls extremely quickly over a distance of 100 meters.

3. Fiber to the Premises

FTTP (fiber to the premises) is a form of fiber-optic communication delivery, in which an optical fiber is run in an optical distribution network from the central office all the way to the premises occupied by the subscriber. The term "FTTP" has become ambiguous and may also refer to FTTC where the fiber terminates at a utility pole without reaching the premises.

Fiber to the premises can be categorized according to where the optical fiber ends:

- FTTH (fiber-to-the-home) is a form of fiber-optic communication delivery that reaches one living or working space. The fiber extends from the central office to the subscriber's living or working space. Once at the subscriber's living or working space, the signal may be conveyed throughout the space using any means, including twisted pair, coaxial cable, wireless, power line communication, or optical fiber.
- FTTB (fiber-to-the-building or -basement) is a form of fiber-optic communication delivery that necessarily applies only to those properties that contain multiple living or working spaces. The optical fiber terminates before actually reaching the subscribers living or working space itself, but does extend to the property containing that living or working space. The signal is conveyed the final distance using any non-optical means, including twisted pair, coaxial cable, wireless, or power line communication.

An apartment building may provide an example of the distinction between FTTH and FTTB. If a fiber is run to a panel inside each subscriber's apartment unit, it is FTTH. If instead the fiber goes only as far as the apartment building's shared electrical room (either only to the ground floor or to each floor), it is FTTB.

4. Fiber to the Curb/Cabinet

Fiber to the curb/cabinet (FTTC) is a telecommunications system based on fiber-optic cables run to a platform that serves several customers. Each of these customers has a connection to this platform via coaxial cable or twisted pair. The "curb" is an abstraction and can just as easily mean a pole-mounted device or communications closet or shed. Typically any system terminating fiber within 1, 000 ft (300 m) of the customer premises equipment would be described as FTTC.

Fiber to the node or neighborhood (FTTN), sometimes identified with and sometimes distinguished from fiber to the cabinet (FTTC), is a telecommunication architecture based on fiber-optic cables run to a cabinet serving a neighborhood. Customers typically connect to this cabinet using traditional coaxial cable or twisted pairwiring. The area served by the cabinet is usually less than one mile in radius and can contain several hundred customers. (If the cabinet serves an area of less than 1, 000 ft (300 m) in radius, the architecture is typically called FTTC/FTTK.)

FTTN allows delivery of broadband services such as high-speed Internet. High-speed communications protocols such as broadband cable access (typically DOCSIS) or some form of digital subscriber line

(DSL) are used between the cabinet and the customers. Data rates vary according to the exact protocol used and according to how close the customer is to the cabinet.

Unlike FTTP, FTTN often uses existing coaxial or twisted-pair infrastructure to provide last mile service and is thus less costly to deploy. In the long term, however, its bandwidth potential is limited relative to implementations that bring the fiber still closer to the subscriber.

A variant of this technique for cable television providers is used in a hybrid fiber-coaxial (HFC) system. It is sometimes given the acronym FTTLA (fiber-to-the-last-amplifier) when it replaces analog amplifiers up to the last one before the customer (or neighborhood of customers).

FTTC allows delivery of broadband services such as high-speed Internet. Usually existing wire is used with communications protocols such as broadband cable access (typically DOCSIS) or some form of DSL connecting the curb/cabinet and the customers. In these protocols, the data rates vary according to the exact protocol used and according to how close the customer is to the cabinet.

Where it is feasible to run new cable, both fiber and copper Ethernet are capable of connecting the "curb" with a full 100Mbit/s or 1Gbit/s connection. Even using relatively cheap outdoor category 5 copper over thousands of feet, all Ethernet protocols including power over Ethernet (PoE) are supported. Most fixed wireless technologies rely on PoE, including Motorola Canopy, which has low-power radios capable of running on a 12VDC power supply fed over several hundred feet of cable.

Power line networking deployments also rely on FTTC. Using the IEEE P1901 protocol (or its predecessor HomePlug AV) existing electric service cables move up to 1Gbit/s from the curb/pole/cabinet into every AC electrical outlet in the home—coverage equivalent to a robust WiFi implementation, with the added advantage of a single cable for power and data.

By avoiding new cable and its cost and liabilities, FTTC costs less to deploy. However, it also has historically had lower bandwidth potential than FTTP. In practice, the relative advantage of fiber depends on the bandwidth available for backhaul, usage-based billing restrictions that prevent full use of last-mile capabilities, and customer premises equipment and maintenance restrictions, and the cost of running fiber that can vary widely with geography and building type.

Historically, both telephone and cable companies avoided hybrid networks using several different transports from their point of presence into customer premises. The increased competitive cost pressure, availability of three different existing wire solutions, smart grid deployment requirements (as in Chattanooga), and better hybrid networking tools (with major vendors like Alcatel-Lucent and Qualcomm Atheros, and WiFi solutions for edge networks, IEEE 1905 and IEEE 802.21 protocols, and SNMP[3] improvements) all make FTTC deployments more likely in areas uneconomic to serve with FTTP/FTTH.

New Words

generic [dʒɪ'nerɪk] *adj.* 类的；一般的；属的；非商标的

widespread ['wɔɪdspred] *adj.* 普遍的，广泛的；分布广的

boundary ['baʊnd(ə)ri] *n.* 边界；范围；分界线

curb	[kɜːb]	*n.* 抑制；路边；勒马绳 *v.* 控制；勒住
Ethernet	[ˈiːθəˌnet]	*n.* [计] 以太网
ambiguously	[æmˈbɪgjʊəsli]	*adv.* 含糊不清地
terminate	[ˈtɜːmɪneɪt]	*v.* 结束，终止；结果
distinguish	[dɪˈstɪŋgwɪʃ]	*v.* 区别，区分；辨别
deploy	[dɪˈplɔɪ]	*n.* 部署 *v.* 配置，展开；部署，展开
acronym	[ˈækrənɪm]	*n.* 首字母缩略词
liability	[laɪəˈbɪlɪti]	*n.* 责任；债务；倾向；可能性；不利因素
backhaul	[ˈbækhɔːl]	*n.* 回程；载货反航
bill	[bɪl]	*v.* 宣布；开账单；用海报宣传
maintenance	[ˈmeɪntənəns]	*n.* 维护，维修；保持；生活费用
uneconomic	[ˌʌniːkəˈnɒmɪk]	*adj.* 不经济的；浪费的

Notes

[1] FTTx 是新一代的光纤用户接入网，用于连接电信运营商和终端用户。FTTx 的网络可以是有源光纤网络，也可以是无源光网络。用于有源光纤网络的成本相对较高，实际上在用户接入网中应用很少，所以目前通常所指的 FFTx 网络应用的都是无源光纤网络。

[2] IEEE（Institute of Electrical and Electronics Engineers）：电气和电子工程师协会是一个国际性的电子技术与信息科学工程师的协会，是目前全球最大的非营利性专业技术学会。IEEE 致力于电气、电子、计算机工程和与科学有关的领域的开发和研究，在太空、计算机、电信、生物医学、电力及消费性电子产品等领域已制定了 900 多个行业标准，现已发展成为具有较大影响力的国际学术组织。

[3] SNMP（简单网络管理协议），由一组网络管理的标准组成，包含一个应用层协议（application layer protocol）、数据库模型（database schema）和一组资源对象。该协议能够支持网络管理系统，用以监测连接到网络上的设备是否有任何需要引起管理上关注的情况。

Questions for Discussion

1. What is the advantage of fiber optic cables over copper cables?
2. How do you understand "fiber is future-proof"?
3. What is the difference between FTTH and FTTB?

Unit 5

Text A

Internet of Things

The Internet of things[1] (stylized Internet of Things or IoT) is the internetworking of physical devices, vehicles (also referred to as "connected devices" and "smart devices"), buildings, and other items—embedded with electronics, software, sensors, actuators, and network connectivity that enable these objects to collect and exchange data. In 2013 the Global Standards Initiative on Internet of Things (IoT-GSI) defined the IoT as "the infrastructure of the information society." The IoT allows objects to be sensed and/or controlled remotely across existing network infrastructure, creating opportunities for more direct integration of the physical world into computer-based systems, and resulting in improved efficiency, accuracy and economic benefit in addition to reduced human intervention. When IoT is augmented with sensors and actuators, the technology becomes an instance of the more general class of cyber-physical systems, which also encompasses technologies such as smart grids, smart homes, intelligent transportation and smart cities. Each thing is uniquely identifiable through its embedded computing system but is able to interoperate within the existing Internet infrastructure. Experts estimate that the IoT will consist of almost 50 billion objects by 2020.

Typically, IoT is expected to offer advanced connectivity of devices, systems, and services that goes beyond machine-to-machine (M2M) communications and covers a variety of protocols, domains, and applications. The interconnection of these embedded devices (including smart objects), is expected to usher in automation in nearly all fields, while also enabling advanced applications like a smart grid, and expanding to the areas such as smart cities.

"Things," in the IoT sense, can refer to a wide variety of devices such as heart monitoring implants, biochip transponders on farm animals, electric clams in coastal waters, automobiles with built-in sensors, DNA analysis devices for environmental/food/pathogen monitoring or field operation devices that assist firefighters in search and rescue operations. Legal scholars suggest looking at "Things" as an "inextricable mixture of hardware, software, data and service". These devices collect useful data with the help of various existing technologies and then autonomously flow the data between other devices. Current market examples include home automation (also known as smart home devices) such as the control and automation of lighting, heating (like smart thermostat), ventilation, air conditioning (HVAC) systems, and appliances such as washer/dryers, robotic vacuums, air purifiers, ovens or refrigerators/freezers that use WiFi for remote monitoring.

As well as the expansion of Internet-connected automation into a plethora of new application areas, IoT is also expected to generate large amounts of data from diverse locations, with the consequent necessity for quick aggregation of the data, and an increase in the need to index, store, and process such data more effectively. IoT is one of the platforms of today's Smart City[2], and Smart Energy Management Systems.

The concept of the IoT was invented by and term coined by Peter T. Lewis in September 1985 in a speech he delivered at a U. S. Federal Communications Commission (FCC)[3] supported session at the Congressional Black Caucus 15th Legislative Weekend Conference.

1. Applications

According to Gartner, Inc. (a technology research and advisory corporation), there will be nearly 20. 8 billion devices on the Internet of things by 2020. ABI Research estimates that more than 30 billion devices will be wirelessly connected to the IoT by 2020. As per a 2014 survey and study done by Pew Research Internet Project, a large majority of the technology experts and engaged Internet users who responded—83 percent—agreed with the notion that the Internet/Cloud of Things, embedded and wearable computing (and the corresponding dynamic systems) will have widespread and beneficial effects by 2025. As such, it is clear that the IoT will consist of a very large number of devices being connected to the Internet. In an active move to accommodate new and emerging technological innovation, the UK Government, in their 2015 budget, allocated £ 40, 000, 000 towards research into the IoT. The former British Chancellor of the Exchequer George Osborne, posited that the IoT is the next stage of the information revolution and referenced the inter-connectivity of everything from urban transport to medical devices to household appliances.

Integration with the Internet implies that devices will use an IP address as a unique identifier. However, due to the limited address space of IPv4 (which allows for 4. 3 billion unique addresses), objects in the IoT will have to use IPv6 to accommodate the extremely large address space required. Objects in the IoT will not only be devices with sensory capabilities, but also provide actuation capabilities (e. g. , bulbs or locks controlled over the Internet). To a large extent, the future of the IoT will not be possible without the support of IPv6; and consequently the global adoption of IPv6 in the coming years will be critical for the successful development of the IoT in the future.

The ability to network embedded devices with limited CPU, memory and power resources means that IoT finds applications in nearly every field. Such systems could be in charge of collecting information in settings ranging from natural ecosystems to buildings and factories, thereby finding applications in fields of environmental sensing and urban planning.

On the other hand, IoT systems could also be responsible for performing actions, not just sensing things. Intelligent shopping systems, for example, could monitor specific users' purchasing habits in a store by tracking their specific mobile phones. These users could then be provided with special offers on their favorite products, or even location of items that they need, which their fridge has automatically conveyed to the phone. Additional examples of sensing and actuating are reflected in applications that deal with heat, electricity and energy management, as well as cruise-assisting transportation systems . Other applications that the IoT can provide is enabling extended home security features and home automation. The concept of an "Internet of living things" has been proposed to describe networks of biological sensors that could use cloud-based analyses to allow users to study DNA or other molecules.

However, the application of the IoT is not only restricted to these areas. Other specialized use cases of the IoT may also exist. An overview of some of the most prominent application areas is provided here. Based on the application domain, IoT products can be classified broadly into five different categories: smart wearable, smart home, smart city, smart environment, and smart enterprise. The IoT products and solutions in each of these markets have different characteristics.

2. Enabling Technologies for IoT

There are many technologies that enable IoT. Crucial to the field is the network used to communicate between devices of an IoT installation, a role that several wireless and/or wired technologies may fulfill:

(1) Short-Range Wireless

- Bluetooth low energy (BLE)—Specification providing a low power variant to classic Bluetooth with a comparable communication range.
- Light-Fidelity (LiFi)—Wireless communication technology similar to the WiFi standard, but using visible light communication for increased bandwidth.
- Near-field communication (NFC)—Communication protocols enabling two electronic devices to communicate within a 4 cm range.
- QR codes and barcodes—Machine-readable optical tags that store information about the item to which they are attached.
- Radio-frequency identification (RFID)—Technology using electromagnetic fields to read data stored in tags embedded in other items.
- Thread—Network protocol based on the IEEE 802.15.4 standard, similar to ZigBee, providing IPv6 addressing.
- WiFi—Widely-used technology for local area networking based on the IEEE 802.11 standard, where devices may communicate through a shared access point.
- WiFi Direct—Variant of the WiFi standard for peer-to-peer communication, eliminating the need for an access point.
- Z-Wave—Communication protocol providing short-range, low-latency data transfer at rates and power consumption lower than WiFi. Used primarily for home automation.
- ZigBee—Communication protocols for personal area networking based on the IEEE 802.15.4 standard, providing low power consumption, low data rate, low cost, and high throughput.

(2) Medium-Range Wireless

• HaLow—Variant of the WiFi standard providing extended range for low-power communication at a lower data rate.

• LTE-Advanced—High-speed communication specification for mobile networks. Provides enhancements to the LTE standard with extended coverage, higher throughput, and lower latency.

(3) Long-Range Wireless

• Low-power wide-area networking (LPWAN)—Wireless networks designed to allow long-range communication at a low data rate, reducing power and cost for transmission.

• Very small aperture terminal (VSAT)—Satellite communication technology using small dish antennas for narrowband and broadband data.

3. Wired

• Ethernet—General purpose networking standard using twisted pair and fiber optic links in conjunction with hubs or switches.

• Multimedia over Coax Alliance (MoCA)—Specification enabling whole-home distribution of high definition video and content over existing coaxial cabling.

• Power-line communication (PLC)—Communication technology using electrical wiring to carry power and data. Specifications such as HomePlug utilize PLC for networking IoT devices.

New Words

stylize	['staɪlaɪz]	*v.* 程式化，风格化
actuator	['æktjʊeɪtə]	*n.* [计] 执行机构；[电]（电磁铁）螺线管
connectivity	[kɒnek'tɪvɪti]	*n.* [数] 连通性；互联互通
intervention	[ˌɪntə'venʃən]	*n.* 介入，干涉，干预，调解
augment	[ɔː g'ment]	*v.* 增强；增加；（使）扩张，扩大
encompass	[ɪn'kʌmpəs]	*v.* 围绕，包围；包含或包括某事物；完成
grid	[grɪd]	*n.* 高压输电线路网；地图坐标方格；格栅
interoperate	[ˌɪntə'ɒpəreɪt]	*n.* /*v.* 交互操作
consist	[kən'sɪst]	*v.* 由……组成；在于；符合
biochip	['baɪəʊtʃɪp]	*n.* 生物芯片
transponder	[træn'spɒndə (r)]	*n.* 发射机应答器，询问机，转发器
pathogen	['pæθədʒ(ə)n]	*n.* 病菌，病原体
inextricable	[ɪn'ekstrɪkəbl]	*adj.* 无法摆脱的；解不开的
plethora	['pleθərə]	*n.* 过多，过剩；[医] 多血，多血症
accommodate	[ə'kɒmədeɪt]	*v.* 容纳；使适应；向……提供住处
corresponding	[ˌkɒrɪs'pɒndɪŋ]	*adj.* 对应的；通信的；符合的；一致的
posit	['pɒzɪt]	*v.* 假定，设想，假设
reference	['refrəns]	*v.* 引用；参照
appliance	[ə'plaɪəns]	*n.* 器具；器械；装置

track	[træk]	*v.* 跟踪；监看，监测
prominent	[ˈprɒmɪnənt]	*adj.* 突出的，杰出的；突起的
twisted	[ˈtwɪstɪd]	*adj.* 扭曲的；喝醉了的；古怪的
specification	[ˌspesɪfɪˈkeɪʃən]	*n.* 规格；详述；说明书

Notes

[1] Internet of things (IoT) 字面意思是"物体组成的互联网"，常译为"物联网"。物联网又称传感网，简要讲就是互联网从人向物的延伸。它是指将各种信息传感设备，如射频识别装置、红外感应器、全球定位系统、激光扫描器等与互联网结合起来而形成的一个巨大网络。其目的是让所有的物品都与网络连接在一起，方便识别和管理。

[2] Smart city"智慧城市"就是运用信息和通信技术手段感测、分析、整合城市运行核心系统的各项关键信息，从而对包括民生、环保、公共安全、城市服务、工商业活动在内的各种需求做出智能响应。其实质是利用先进的信息技术，实现城市智慧式管理和运行，进而为城市中的人创造更美好的生活，促进城市的和谐、可持续发展。

[3] Federal Communications Commission (FCC) 是指美国联邦通信委员会。该机构直接对国会负责，通过控制无线电广播、电视、电信、卫星和电缆来协调国内和国际的通信，负责授权和管理除联邦政府使用之外的射频传输装置和设备。

Questions for Discussion

1. What does the Internet of things mean?
2. What do you think of the future of IoT?
3. How many types can medium-range wireless fall into? What are they?

Text B

Forget the Internet of Things: Here Comes the "Internet of Cars"

Figure 5-1

What if large groups of people could go beyond ridesharing—replacing traditional car ownership altogether through on-demand access to the cars they want: A convertible in the summer, an SUV for winter ski trips?

What if driving skills could be computed as a score that warned us of bad drivers nearby—real time, on the road—also enabling navigation systems to offer safer alternative routes? Imagine if we could get rid of traffic jams and accidents altogether. Or how about if our cars picked up our groceries on their own—and dropped us off at the airport like a self-contained limo service?

What if automakers could subsidize[1] our car purchases by working with telecommunications and other companies that want to capitalize on the lifetime revenue opportunity of a connected driver? Consider also the possibilities for insurance providers to charge higher premiums (for those who drive their cars themselves), or for local governments to monitor personal CO_2 usage (in exchange for not taxing or tolling public roads).

Whether you embrace or object to these scenarios, they're not too far away. This isn't just an evolution of technology-enabled, connected vehicles. This goes beyond self-driving cars. And it's more than a simple sensor-network: This is the era of smart mobility—an Internet of Cars.

Basically, cars have become the "ultimate mobile device" and we, the people, are becoming "connected drivers". These aren't just buzzwords: As a longtime strategic adviser and analyst of this space, I've been using these terms since 1998 to describe this fundamental transformation of the automobile. And it's coming within this decade. For example, by 2016, most buyers in mature automotive markets (U. S., Western Europe) will consider vehicles' ability to access web-based information a key criterion when purchasing an automobile. For premium vehicle brand buyers, this tipping point will be reached even sooner: 2014. That's just one year away.

The connected vehicle is leading the automotive industry to its most significant innovation phase... since the creation of the automobile itself.

1. The Era of Smart Mobility Is Going to Change Everything

But what is it? "Connected vehicles" are cars that access, consume, create, enrich, direct, and share digital information between businesses, people, organizations, infrastructures, and things. Those "things" include other vehicles, which is where the Internet of Things becomes the Internet of Cars.

As these vehicles become increasingly connected, they become self-aware, contextual, and eventually, autonomous. Those of you reading this will probably experience self-driving cars in your lifetime—though maybe not all three of its evolutionary phases: from automated to autonomous to unmanned.

We still need to address a number of technology, engineering, legislative, and market issues to develop successful offerings here. But this automotive era builds on current and related industry trends such as the convergence of digital lifestyles, the emergence of new mobility solutions, demographic shifts, and the rise of smartphones and the mobile Internet.

Consumers now expect to access relevant information wherever they are... including in the automobile. At the same time, these technologies are making new mobility solutions—such as peer-to-peer car sharing—more widespread and attractive. This is especially important since vehicle ownership in urban areas is expensive and consumers, especially younger ones, don't show the same desire for vehicle ownership as older generations do.

To be successful, connected vehicles will draw on the leading technologies in sensors, displays, on-board and off-board computing, in-vehicle operating systems, wireless and in-vehicle data communication, machine learning, analytics, speech recognition, and content management. (That's just to name a few.) All of this leads to considerable benefits and opportunities: reduced accident rates, increased productivity, improved traffic flow, lowered emissions, extended utility for EVs[2], new entertainment options, and new marketing and commerce experiences.

Besides providing automobiles and drivers with new function, connected vehicles will also expand automotive business models to include a much broader set of industries—IT, retail, financial services, media, and consumer electronics. This is significant, because it could challenge the traditional automotive business model: Rather than focusing only on the sale and maintenance of a vehicle, companies will focus on the sum of business opportunities the automobile represents.

2. But What Do Consumers Want?

Do people even want all this? Or is this just a case of business thinkers, technologists, and early adopters making predictions in an echo chamber? It's not. Consumers do show a strong interest in the features of a connected vehicle. For example, from analyses Gartner conducted over the last year, we found that of all U. S. vehicle owners:

- Almost half (46%) are interested in safely accessing mobile applications inside the vehicle. These applications include receiving on-demand wireless map or software updates, finding available parking spots, and conducting local searches; nearly 40% would also opt for remote diagnostic capabilities that alert them when parts need replacement.
- More than one-third are interested in a self-driving, autonomous vehicle.
- Thirty percent are likely to opt for a vehicle that allows them to tether their smartphone to get Internet connection there.

Our increasingly digital "lifestyles" may also force consumers to re-evaluate personal transportation choices. For example: The combined cost of a monthly mobile and residential Internet plan might be competing with the cost of filling up a car at the gas station.

These tradeoffs are even more important to younger vehicle owners (18-to 24-year-olds) than older ones (54 + years). The younger group is more likely (30%) to choose Internet access over having a vehicle (compared to just 12% of the older group), and about the same percentages are likely to use a car-sharing service as an alternative to vehicle ownership.

Obviously, connected vehicle applications have to be safe, reliable, and non-distracting to wow consumers on an emotional level and convince them on a rational level. Simply copying interfaces from other mobile devices will not be enough—buttons in cars actually work great for certain functions. The automotive industry will need to innovate new experiences and integrate systems thoroughly so consumers don't feel they can get the same results with just an iPad on the passenger seat.

But the fact remains that automobiles are here to stay, and they're going to be connected. The innovations and changes described here will mature relatively quickly over the next two decades. For example, I predict that by 2016 at least three companies will have announced concrete plans for upcoming product launches offering advanced autonomous vehicle technology.

This isn't pie-in-the-sky—just consider a few recent advancements in the automotive connectivity space: Avis acquiring Zipcar[3]; the first over-the-air automotive software patch by Tesla; Intel getting significantly involved in the connected vehicle value-chain; big telcos like Sprint extending their reach into automotive; a high-ranking Apple executive taking a seat on a carmaker's board. All of these moves signal the trend.

And for those who are also passionate about automobiles and driving, the era of the connected vehicle will open a mesmerizing new world. You know that immediate connection between our senses and the stimulatory triggers of a car: sounds, speed, sights? Imagine that feeling, and so much more. I am optimistic that the automotive industry and technology companies will preserve this fascination of the automobile—it is, after all, an immersive experience.

But if you don't like this dawning era of the connected vehicle, you should get your (unconnected) dream car now.

(By *Thilo Koslowski*[4] from *Wired*)

New Words

rideshare	[ˈraɪdʃeə]	*n.* 共同乘车（以减少费用）
limo	[ˈlɪməʊ]	*n.* 豪华轿车
subsidize	[ˈsʌbsɪdaɪz]	*v.* 资助；给予奖助金；向……行贿
scenario	[sɪˈnɑːrɪəʊ]	*n.* 方案；情节；剧本；设想
sensor	[ˈsensə]	*n.* 传感器
strategic	[strəˈtiːdʒɪk]	*adj.* 战略上的，战略的
contextual	[kɒnˈtekstjʊəl]	*adj.* 上下文的；前后关系的
legislative	[ˈledʒɪslətɪv]	*n.* 立法权；立法机构 *adj.* 立法的；有立法权的
utility	[juːˈtɪlɪti]	*n.* 实用；效用；公共设施
chamber	[ˈtʃeɪmbə]	*n.* （身体或器官内的）室；房间；会所
opt	[ɒpt]	*v.* 选择
tradeoff	[ˈtreɪd, ɔːf]	*n.* 权衡；折中；（公平）交易（等于 trade-off）
trend	[trend]	*n.* 趋势，倾向；走向
mesmerize	[ˈmezməraɪz]	*v.* 施催眠术；迷住；以魅力迫使
immersive	[ɪˈmɜːsɪv]	*adj.* 拟真的；沉浸式的

Notes

[1] subsidize 意为"以津贴补助"，即购车补贴。

[2] EVs: electric vehicles "电动汽车"。

[3] Zipcar 是美国的一家分时租赁互联网汽车共享平台。该平台由罗宾·蔡斯（Robin

Chase）与安特耶·丹尼尔斯（Antje Danielson）于2000年共同创办。Zipcar主要以“汽车共享”为理念，其汽车停放在居民集中地区，会员可以通过网站、电话和应用软件搜寻需要的车辆，选择就近预约取车和还车，车辆的开启和锁停完全通过一张会员卡完成。

[4] Thilo Koslowski是市场分析公司Gartner的分析师。他认为汽车将成为终极移动设备，苹果和谷歌都在通过各项合作将自家的技术整合到汽车中。

Questions for Discussion

1. What are connected cars?
2. What are the benefits and opportunities of connected cars?
3. What is the demand of U. S. vehicle owners according to the analyses Gartner conducted?

Unit 6

Text A

Quantum Communications Leap Out of the Lab

——China begins work on super-secure network as "real-world" trial successfully sends quantum keys and data.

Cybersecurity is a step closer to the dream of sending data securely over long distances using quantum physics—spurred by two developments.

This week, China will start installing the world's longest quantum-communications network, which includes a 2, 000-kilometre link between Beijing and Shanghai. And a study jointly announced this week by the companies Toshiba, BT and ADVA, with the UK National Physical Laboratory in Teddington, reports "encouraging" results from a network field trial, suggesting that quantum communications could be feasible on existing fibre-optic infrastructure.

Is there any disadvantage in conventional data-encryption systems? What is it?

Conventional data-encryption systems rely on the exchange of a secret "key" —in binary 0s and 1s—to encrypt and decrypt information. But the security of such a communication channel can be undermined if a hacker "eavesdrops" on this key during transmission. Quantum communications use a technology called quantum key distribution (QKD)[1], which harnesses the subatomic properties of photons to "remove this weakest link of the current system", says Grégoire Ribordy, co-founder and chief executive of ID Quantique, a quantum-cryptography company in Geneva, Switzerland.

The method allows a user to send a pulse of photons that are placed in specific quantum states that characterize the cryptographic key. If anyone tries to intercept the key, the act of eavesdropping intrinsically alters its quantum state—alerting users to a security breach. Both the US $ 100-million Chinese initiative and the system tested in the latest study use QKD.

The Chinese network "will not only provide the highest level of protection for government and financial data, but provide a test bed for quantum theories and new technologies", says Jianwei Pan, a quantum physicist at the University of Science and Technology of China in Hefei, who is leading the Chinese project.

Pan hopes to test such ideas using the network, along with a quantum satellite that his team plans to launch next year. Together, he says, the technologies could perform further tests of fundamental quantum theories over large scales (around 2, 000 kilometres), such as quantum non-locality[2], in which changing the quantum state of one particle can influence the state of another even if they are far apart, says Pan.

Sending single photons over long distances is one of the greatest problems in QKD because they tend to get absorbed by optical fibers, making the keys tricky to detect on the receiver's end.

This is "a big challenge for conventional detectors", says Hoi-Kwong Lo, a quantum physicist at the University of Toronto in Canada. But technological breakthroughs in recent years have significantly reduced the noise level of detectors while increasing their efficiency in detecting photons from just 15% to 50%.

Vast improvements have also been made in the rate at which detectors can "count" photon pulses—crucial in determining the rate at which quantum keys can be sent, and thus the speed of the network. Counting rates have been raised 1, 000-fold, to about 2 gigahertz, says Lo.

The breakthroughs are pushing the distance over which quantum signals can be sent. Trials using "dark fibers" —optical fibers laid down by telecommunications companies but lying unused—have sent quantum signals up to 100 kilometers, says Don Hayford, a researcher at Battelle, a technology-development company headquartered in Columbus, Ohio.

To go farther than that, quantum signals must be relayed at "node points" —the quantum networks between Beijing and Shanghai, for instance, will require 32 nodes. To transmit photons over longer distances without the use of nodes would require a satellite.

China is not alone in its quantum-communication efforts. A team led by Hayford, together with ID Quantique, has started installing a 650-kilometre link between Battelle's headquarters and its offices in Washington D C. The partnership is also planning a network linking major U. S. cities, which could exceed 10, 000 kilometers, says Hayford, although it has yet to secure funding for that.

The Chinese and U. S. networks will both use dark fibers to send quantum keys. But these fibers "are not always available and can be prohibitively expensive", says Andrew Shields, a quantum physicist at Toshiba Research Europe in Cambridge, UK. One way to sidestep the problem is to piggyback the photon streams onto the "lit" fibers that transmit conventional telecommunications data. However, those conventional data streams are usually about a million times stronger than quantum streams, so the quantum data tend to be drowned out.

In the results announced this week, Shields and his colleagues were successful in achieving the stable and secure transmission of QKDs along a live lit fiber between two stations of the UK telecommunications company BT, 26 kilometers apart. The quantum keys were sent over several weeks at a high rate alongside four channels of strong conventional data on the same fiber. The research builds on previous work in which Shields and his team developed a technique to detect quantum signals sent alongside noisy data in a 90-kilometre fiber, but in controlled laboratory conditions.

"Implementing QKD in the 'real world' is much more challenging than in the controlled environment of the lab, due to environmental fluctuations and greater loss in the fibre," says Shields.

The quantum keys in the latest study were sent alongside conventional data travelling at 40 gigabits per second. "As far as I am aware, this is the highest bandwidth of data that has been multiplexed with QKD to date," add Shields.

He calculates that it would be possible to send QKD signals alongside 40 conventional data channels. Optical fibers usually carry between 40 and 160 telecommunications channels, meaning that quantum communication could be carried out with existing infrastructure.

"I find it an impressive piece of work that demonstrates the multiplexing of strong classical signals with quantum signals in the same fiber for the first time" in a field trial, says Lo. Removing the need for dark fibers, he says, is an important step in showing that QKD has the potential to be used in "real life".

By *Jane Qiu*, 23 April 2014, from *Nature*

New Words

spur	[spɜː]	*v.* 促进，加速，推动
feasible	[ˈfiːzəbl]	*adj.* 可行的；可用的；可实行的
conventional	[kənˈvenʃənl]	*adj.* 传统的；平常的；依照惯例的；约定的
encrypt	[ɪnˈkrɪpt]	*v.* 编密码；加密
decrypt	[dɪˈkrɪpt]	*v.* 译（电文），解释明白，解密
undermine	[ˌʌndəˈmaɪn]	*v.* 逐渐削弱；使逐步减少效力；破坏
harness	[ˈhaːnɪs]	*v.* 利用；控制；给（马等）套轭具
intercept	[ˌɪntəˈsept]	*v./n.* 拦截，拦住；截球；截击
subatomic	[ˈsʌbəˈtɒmɪk]	*adj.* 小于原子的，亚原子的，次原子的
cryptography	[krɪpˈtɒgrəfi]	*n.* 密码使用法，密码系统，密码术
photon	[ˈfəʊtɒn]	*n.* <物>光子，光量子
eavesdrop	[ˈiːvzdrɒp]	*v.* 偷听；窃听
intrinsic	[ɪnˈtrɪnsɪk]	*adj.* 固有的，内在的，本质的
breach	[briːtʃ]	*n.* 破坏；破裂；缺口
initiative	[ɪˈnɪʃɪətɪv]	*n.* 积极的行动；倡议；主动
locality	[ləʊˈkælɪti]	*n.* 位置，地区；产地
tricky	[ˈtrɪki]	*adj.* 复杂的；棘手的；微妙的
gigahertz	[ˈgɪgəhɜːts]	*n.* 千兆赫
prohibitive	[prəˈhɪbɪtɪv]	*adj.* 禁止的，禁止性的；抑制的
sidestep	[ˈsaɪdstep]	*v.* 回避，绕开；向侧方跨步

piggyback	[ˈpɪgɪbæk]	*v.* 背负式装运
fluctuation	[ˌflʌktjʊˈeɪʃən]	*n.* 波动，涨落，起伏
multiplex	[ˈmʌltɪpleks]	*v.* ［通信］多路传输
spur	[spɜː]	*v.* 促进，加速，推动
feasible	[ˈfiːzəbl]	*adj.* 可行的；可用的；可实行的
conventional	[kənˈvenʃənl]	*adj.* 传统的；平常的；依照惯例的；约定的

Notes

[1] quantum key distribution (QKD)：量子密钥分配是1984年物理学家查尔斯H. 贝内特（Charles H. Bennett）和密码学家吉勒斯·布拉萨德（Gilles Brassard）提出的基于量子力学测量原理的BB84协议，量子密钥分配从根本上保证了密钥的安全性。例如，天宫二号上的载荷“量子密钥分配专项”就是以实现空地间实用化的量子密钥分配为目标，通过天上发射一个个单光子并在地面接收，生成“天机不可泄露”的量子密钥。

[2] quantum non-locality：量子的非定域性又称“测不准原理”“不确定关系”，是量子力学的一个基本原理，由德国物理学家沃纳·卡尔·海森堡（Werner Karl Heisenberg）于1927年提出。该原理表明，一个微观粒子的某些物理量（如位置和动量，或方位角与动量矩，还有时间和能量等），不可能同时具有确定的数值，其中一个量越确定，另一个量的不确定程度就越大。

Questions for Discussion

1. Is there any disadvantages in conventional data-encryption systems? What is it?
2. How can QKD prevent hackers from eavesdropping during transmission?
3. What is the disadvantage of dark fibers? How can it be avoided?
4. Do you think it is possible that QKD will be employed in “real life”?

Text B

China's Latest Leap Forward Isn't Just Great—It's Quantum

——Beijing launches the world's first quantum-communications satellite into orbit

BEIJING—A rocket that shot skyward from the Gobi Desert early Tuesday is expected to propel China to the forefront of one of science's most challenging fields.

It also is set to launch Beijing far ahead of its global rivals in the drive to acquire a highly coveted asset in the age of cyberespionage: hack-proof communications.

State media said China sent the world's first quantum-communications satellite into orbit from a launch center in Inner Mongolia about 1:40 a. m. Tuesday. Five years in the making, the project is being closely watched in scientific and security circles.

The quantum program is the latest part of China's multibillion-dollar strategy over the past two decades to draw with or even surpass the West in hard-sciences[1] research.

"There's been a race to produce a quantum satellite, and it is very likely that China is going to win that race," said Nicolas Gisin, a professor and quantum physicist at the University of Geneva. "It shows again China's ability to commit to large and ambitious projects and to realize them."

Scientists in the U. S., Europe, Japan and elsewhere are rushing to exploit the strange and potentially powerful properties of subatomic particles, but few with as much state support as those in China, researchers say. Quantum technology is a top strategic focus in the country's five-year economic development plan, released in March.

Beijing hasn't disclosed how much money it has allocated to quantum research or to building the 1,400-pound satellite. But funding for basic research, which includes quantum physics, was $10.1 billion in 2015, up from $1.9 billion in 2005.

U. S. federal funding for quantum research is about $200 million a year, according to a congressional report in July by a group of science, defense, intelligence and other officials. It said development of quantum science would "enhance U. S. national security," but said fluctuations in funding had set back progress.

Beijing, meanwhile, has tried to lure Chinese-born, foreign-educated experts in quantum physics back to China, including Pan Jianwei, the physicist who is leading the project.

"We've taken all the good technology from labs around the world, absorbed it and brought it back," Mr. Pan told Chinese state TV in an interview that aired on Monday.

With state support, Mr. Pan was able to leapfrog his former Ph. D. adviser, University of Vienna physicist Anton Zeilinger, who said he has tried since 2001 to persuade the European Space Agency to launch a similar satellite.

"It's a difficult process, which takes a lot of time," said Mr. Zeilinger, who is now working on his former student's satellite.

Neither Mr. Pan nor the Chinese Academy of Sciences, which is directing the project, responded to requests to comment. The European Space Agency didn't respond to requests for comment.

The National Science Foundation, which provides federal funding for basic American science research, said quantum science was one of six "big ideas." It had identified for long-term research to address critical societal challenges, according to C. Denise Caldwell, head of the group's physics division. She said the NSF also recently established new research awards that will have a direct impact on quantum information research.

China's investment in the field is likely being driven in part by fear of U. S. cyber capabilities, said John Costello, a fellow at Washington, D. C. -based New America specializing in China and cybersecurity, pointing to 2013 disclosures that the U. S. had penetrated deeply into Chinese networks. He also noted that U. S. institutions are researching how to build powerful quantum computers theoretically capable of shattering the math-based encryption now used world-wide for secure communication.

"The Chinese government is aware that they are growing particularly susceptible to electronic espionage," Mr. Costello said.

However, quantum communication is defensive in nature, he noted, and wouldn't benefit from what the U. S. has identified as China's state-sponsored hacking program.

Quantum encryption is secure against any kind of computing power because information encoded in a quantum particle is destroyed as soon as it is measured. Gregoir Ribordy, co-founder of Geneva-based quantum cryptography firm ID Quantique, likened it to sending a message written on a soap bubble. "If someone tries to intercept it when it's being transmitted, by touching it, they make it burst," he said.

Quantum physicists have recently advanced the use of photons to communicate securely over short distances on Earth. The satellite, if successful, would vastly expand the range of un-hackable communication. To test whether quantum communications can take place at a global scale, Mr. Pan has told state media that he and his team will try to beam a quantum cryptographic key through space from Beijing to Vienna. "Inevitably these kinds of technologies have problems and things get messed up by the people using them, unless they have gone through extensive training," said Peter Mattis, a fellow at the Jamestown Foundation[2] who studies China's intelligence services. "I think China has an obligation not just to do something for ourselves—many other countries have been to the moon, have done manned spaceflight—but to explore something unknown," he said.

"It would be enormous if the test succeeded," said Ma Xiaosong, a Vienna-trained quantum physicist at Nanjing University who worked on early phases of the satellite project.

Quantum encryption isn't foolproof. It's possible for hackers to trick an incautious recipient by shining an intense laser into a quantum receptor, said Alexander Ling, principal investigator at the Center for Quantum Technologies in Singapore.

U. S. security experts also question whether intricacies of quantum communication can be simplified enough for use in a conflict situation.

Whatever the challenges, the University of Vienna's Mr. Zeilinger said, the satellite puts China and the field of quantum mechanics on the verge of a significant technological breakthrough. "In the long run, there is a good chance that this will replace our current communications technology," he said. "I see no basic reason why it won't happen."

In a January interview with the journal *Nature*, Mr. Pan said the satellite showed China's scientists had stopped following in the footsteps of others. To drive the point home, Chinese state media on Monday said the satellite had been named Micius after a 5th century B. C. philosopher who opposed offensive warfare.

By *JOSH CHIN* from *Wall Street Journal*, *Vivian Pang* contributed to this article.

New Words

orbit	[ˈɔːbɪt]	*n.* 轨道；生活常规 *v.* 盘旋；绕轨道运行
skyward	[ˈskaɪwəd]	*adj.* 向上的；向着天空的 *adv.* 向上；朝天空
propel	[prəˈpel]	*v.* 推进；驱使；激励；驱策
forefront	[ˈfɔːfrʌnt]	*n.* 最前线，最前部；活动的中心

covet	[ˈkʌvɪt]	*v.* 垂涎；渴望；贪图
cyberespionage	[ˈsaɪbəˈespɪənaːʒ]	*n.* 网上间谍活动
hack-proof	[hæk pruːf]	*n.* 防黑客
lure	[l (j) ʊə]	*n.* 诱惑；饵；诱惑物 *v.* 诱惑；引诱
leapfrog	[ˈliːpfrɒg]	*n.* 交互跃进 *v.* 跳背，跳蛙，交替前进；跃过
penetrate	[ˈpenɪtreɪt]	*v.* 渗透，穿透，洞察；渗透，刺入，看透
susceptible	[səˈseptɪb(ə)l]	*adj.* 易受影响的；易感动的 *n.* 易得病的人
foolproof	[ˈfuːlpruːf]	*adj.* 十分简单的；不会错的 *n.* 极简单；安全自锁装置
incautious	[ɪnˈkɔːʃəs]	*adj.* 不小心的；轻率的
offensive	[əˈfensɪv]	*adj.* 攻击的；冒犯的；无礼的；讨厌的
warfare	[ˈwɔːfeə]	*n.* 战争；冲突

Notes

[1] Hard science（硬科学）是自然科学与技术科学交叉的统称，内容是数学、物理学、化学、天文学等。

[2] Jamestown Foundation（詹姆斯顿基金会）是美国的一个有关军事或战争的基金会。这个基金会是一个专家用于研究军事科技的专项基金会。

Questions for Discussion

1. What project has received close attention in scientific and security circles for the past five years?
2. How do you understand "quantum encryption is secure"?
3. Why is quantum encryption not foolproof?

Unit 7

Text A

Stored Program Control

Stored program control (SPC) is a telecommunications technology used for telephone exchanges controlled by a computer program stored in the memory of the switching system. SPC was the enabling technology of electronic switching systems (ESS) developed in the Bell System[1] in the 1950s.

Early exchanges such as Strowger, panel, rotary, and crossbar switches were constructed purely from electromechanical switching components with analog control electronics, and had no computer software control. Stored program control was invented by Bell Labs[2] scientist Erna Schneider Hoover in 1954 who reasoned that computer software could control the connection of telephone calls.

SPC was introduced in electronic switching systems in the 1960s. The 101ESS PBX was a transitional switching system in the Bell System to provide expanded services to business customers that were otherwise still served by an electromechanical central office switch, while the Western Electric 1ESS switch and the AXE telephone exchange by Ericsson were large-scale systems in the public switched telephone network. SPC enabled sophisticated calling features. As SPC exchanges evolved, reliability and versatility increased. The addition of time division multiplexing (TDM) decreased subsystem sizes and dramatically increased the capacity of the telephone network. By the 1980s, SPC technology dominated the telecommunications industry.

1. Introduction

The principle feature of stored program control is one or multiple digital processing units (stored program computers) that execute a set of computer instructions (program) stored in the memory of the system by which telephone connections are established, maintained, and terminated in associated electronic circuitry.

An immediate consequence of stored program control is automation of exchange functions and introduction of a variety of new telephony features to subscribers.

A telephone exchange must run continuously without interruption at all times, by implementing a fault-tolerant design. Early trials of electronics and computers in the control sub systems of an exchange were successful and resulted in the development of fully electronic systems, in which the switching network was also electronic. A trial system with stored program control was installed in Morris, Illinois in 1960. It used a flying-spot store with a word size of 18 bits for semi-permanent program and parameter storage, and a barrier-grid memory for random access working memory. The world's first electronic switching system for permanent production use, the No. 1 ESS, was commissioned by AT&T at Succasunna, New Jersey, in May 1965. By 1974, AT&T had installed 475 No. 1ESS systems. In the 1980s SPC displaced electromechanical switching in the telecommunication industry, hence the term lost all but historical interest. Today SPC is a standard feature in all electronic exchanges.

The attempts to replace the electromechanical switching matrices by semiconductor cross point switches were not immediately successful, particularly in large exchanges. As a result, many space division switching systems used electromechanical switching networks with SPC. Nonetheless, private automatic branch exchanges (PABX) and smaller exchanges do use electronic switching devices.

2. Types

Stored program control implementations may be organized into centralized and distributed approaches. Early electronic switching systems (ESS) developed in the 1960s and 1970s almost invariably used centralized control. Although many present day exchange design continue to use centralized SPC, with advent of low cost powerful microprocessors and VLSI chips such as programmable logic array (PLA) and programmable logic controllers (PLC), distributed SPC became widespread by the early 21st century.

(1) Centralized Control

In centralized control, all control equipment is replaced by a central processing unit. It must be able to process 10 to 100 calls per second, depending on the load to the system. Multiprocessor configurations are commonplace and may operate in various modes, such as in stand-by mode, in synchronous duplex mode, or in load-sharing mode.

1) Stand-by Mode

Standby mode of operation is the simplest of a dual-processor configuration. Normally one processor is in stand-by mode. The stand-by processor is brought online only when the active processor fails. An important requirement of this configuration is ability of stand-by processor to reconstitute the state of exchange system when it takes over the control; means to determine which of the subscriber lines or trunks are in use.

In small exchanges, this may be possible by scanning the status signals as soon as the stand-by processor is brought into action. In such a case only the calls which are being established at the time of failure are disturbed. In large exchanges it is not possible to scan all the status signals within a significant time. Here the active processor copies the status of system periodically into secondary storage. When switchover occurs the recent status from the secondary memory is loaded. In this case only the calls which change status between last update and failure are affected. The shared secondary storage need not to be duplicated and simple unit level redundancy would suffice. 1ESS switch was a prominent example.

2) Synchronous Duplex Mode. In synchronous duplex mode of operation hardware coupling is provided between two processors which execute same set of instructions and compare the results continuously. If mismatch occurs then the faulty processor is identified and taken out of service within a few milliseconds. When system is operating normally, the two processors have same data in memories at all times and simultaneously receive information from exchange environment. One of the processor actually controls the exchange, but the other is synchronized with the former but does not participate in the exchange control. If a fault is detected by the comparator the processors are decoupled and a check-out program is run independently to find faulty processor. This process runs without disturbing the call processing which is suspended temporarily. When one processor is taken out then the other processor operates independently. When the faulty processor is repaired and brought in service then memory contents of the active processor are copied into its memory and the two are synchronized and comparator is enabled.

It is possible that a comparator fault occurs only due to transient failure which is not shown even when check out program is run. In such case three possibilities exists:

- Continue with both processors;
- Take out the active processor and continue with the other;
- Continue with active processor but remove other processor from service.

Scheme 1 is based on the assumption that the fault is transient one and may not appear again. In scheme 2 and Scheme 3 the processor taken out is subjected to extensive testing to identify a marginal failure in these cases.

3) Load-Sharing Mode. In load-sharing operation, an incoming call is assigned randomly or in a predetermined order to one of the processors which then handles the call right through completion. Thus, both the processors are active simultaneously and share the load and the resources dynamically. Both the processors have access to the entire exchange environment which is sensed as well as controlled by these processors. Since the calls are handled independently by the processors, they have separate memories for storing temporary call data. Although programs and semi-permanent data can be shared, they are kept in separate memories for redundancy purposes.

There is an inter processor link through which the processors exchange information needed for mutual coordination and verifying the "state of health" of the other. If the exchange of information fails, one of the processors which detect the same takes over the entire load including the calls that are already set up by the failing processor. However, the calls that were being established by the failing processor are usually lost. Sharing of resources calls for an exclusion mechanism so that both the processors do not seek the same resource at the same time. The mechanism may be implemented in software or hardware or both.

(2) Distributed Control

Distributed SPC is both more available and more reliable than centralized SSPC.

1) Vertical Decomposition. Whole exchange is divided into several blocks and a processor is assigned to each block. This processor performs all the task related to that specific blocks. Therefore, the total control system consists of several controlunits coupled together. For redundancy purpose processor may be duplicated in each blocks.

2) Horizontal Decomposition. In this type of decomposition each processor performs only one or some exchange function.

New Words

memory	['mem(ə)ri]	*n.* [计] 存储器，内存；记忆
panel	['pæn(ə)l]	*n.* 仪表板；嵌板
rotary	['rəʊt(ə)ri]	*adj.* 旋转的，转动的 *n.* [动力] 转缸式发动机
crossbar	['krɒsbɑː]	*n.* 横木；横条；横梁，横档，闩
electromechanical	[ɪˌlektrəʊmɪ'kænɪk(ə)l]	*adj.* [电] [机] 电动机械的，[机] 机电的
transitional	[træn'zɪʃ(ə)n(ə)l]	*adj.* 渐变的，转变的；变迁的，过渡期的
versatility	[ˌvɜːsə'tɪləti]	*n.* 多用途；可转动性；多才多艺
execute	['eksɪkjuːt]	*v.* 执行；处死，处决；履行；完成
parameter	[pə'ræmɪtə]	*n.* 参数；参量；限制因素；决定因素
barrier-grid	['bærɪə (r) grɪd]	*n.* 障栅
commission	[kə'mɪʃ(ə)n]	*v.* 委任，授予；使服役
displace	[dɪs'pleɪs]	*v.* 移动，移走；替换，取代
programmable	[ˌprəʊ'græməbl]	*adj.* 可设计的，可编程的；可编程序
duplex	['djuːpleks]	*adj.* 有两部分的
standby	['stæn (d) baɪ]	*n.* 备用品 *adj.* 备用的
reconstitute	[riː'kɒnstɪtjuːt]	*v.* 再组成，再构成
suffice	[sə'faɪs]	*v.* 足够；有能力
duplicate	['djuːplɪkeɪt]	*v.* 复制；重复 *adj.* 完全一样的；复制的
redundancy	[rɪ'dʌnd(ə)nsi]	*n.* （机器的）多余度；过多；冗长；裁员
synchronous	['sɪŋkrənəs]	*adj.* 同时存在 [发生] 的，同步的
synchronize	['sɪŋkrənaɪz]	*v.* 使同步；使同时
comparator	[kəm'pærətə]	*n.* 比较仪，比较器；比测器
transient	['trænzɪənt]	*adj.* 短暂的；转瞬即逝的；临时的
predetermine	[priːdɪ'tɜːmɪn]	*v.* 预先裁定；注定
couple	['kʌp(ə)l]	*v.* 连在一起，连接
decomposition	[ˌdiːkɒmpə'zɪʃn]	*n.* 分解；腐烂

Notes

[1] Bell System：贝尔本人创立的贝尔电话公司曾形成庞大的贝尔系统，并垄断美国的电信

事业达百年之久，贝尔系统以 AT&T 公司为母公司，下属众多子公司和研究所。

[2] Bell Labs：美国贝尔实验室是晶体管、激光器、太阳电池、发光二极管、数字交换机、通信卫星、电子数字计算机、蜂窝移动通信设备、长途电视传送、仿真语言、有声电影、立体声录音以及通信网等许多重大发明的诞生地。自 1925 年以来，贝尔实验室共获得 25000 多项专利，现在，平均每个工作日获得三项多专利。

Questions for Discussion

1. What is stored program control?
2. What main feature does SPC bear?
3. How many categories does SPC fall into? What are they?
4. What is stand-by mode?

Text B

Packet Switching

Packet switching is a digital networking communications method that groups all transmitted data into suitably sized blocks, called packets, which are transmitted via a medium that may be shared by multiple simultaneous communication sessions. Packet switching increases network efficiency, robustness and enables technological convergence of many applications operating on the same network.

Packets are composed of a header and payload. Information in the header is used by networking hardware to direct the packet to its destination where the payload is extracted and used by application software.

Starting in the late 1950s, American computer scientist Paul Baran[1] developed the concept Distributed Adaptive Message Block Switching with the goal to provide a fault-tolerant, efficient routing method for telecommunication messages as part of a research program at the RAND Corporation[2], funded by the U. S . Department of Defense. This concept contrasted and contradicted then established principles of pre-allocation of network bandwidth, largely fortified by the development of telecommunications in the Bell System. The new concept found little resonance among network implementers until the independent work of British computer scientist Donald Davies[3] at the National Physical Laboratory (United Kingdom) in the late 1960s. Davies is credited with coining the modern name packet switching and inspiring numerous packet switching networks in the decade following, including the incorporation of the concept in the early ARPANET in the United States.

1. Concept

A simple definition of packet switching is:

The routing and transferring of data by means of addressed packets so that a channel is occupied during the transmission of the packet only, and upon completion of the transmission the channel is made available for the transfer of other traffic.

Packet switching features delivery of variable bit rate data streams, realized as sequences of packets, over a computer network which allocates transmission resources as needed using statistical multiplexing or dynamic bandwidth allocation techniques. As they traverse network nodes, such as switches and routers, packets are received, buffered, queued, and transmitted (stored and forwarded), resulting in variable latency and throughput depending on the link capacity and the traffic load on the network.

Packet switching contrasts with another principal networking paradigm, circuit switching, a method which pre-allocates dedicated network bandwidth specifically for each communication session, each having a constant bit rate and latency between nodes. In cases of billable services, such as cellular communication services, circuit switching is characterized by a fee per unit of connection time, even when no data is transferred, while packet switching may be characterized by a fee per unit of information transmitted, such as characters, packets, or messages.

Packet mode communication may be implemented with or without intermediate forwarding nodes (packet switches or routers). Packets are normally forwarded by intermediate network nodes asynchronously using first-in, first-out buffering, but may be forwarded according to some scheduling discipline for fair queuing, traffic shaping, or for differentiated or guaranteed quality of service, such as weighted fair queuing or leaky bucket. In case of a shared physical medium (such as radio or 10BASE5), the packets may be delivered according to a multiple access scheme.

2. History

In the late 1950s, the US Air Force established a wide area network for the Semi-Automatic Ground Environment (SAGE) radar defense system. They sought a system that might survive a nuclear attack to enable a response, thus diminishing the attractiveness of the first strike advantage by enemies.

Leonard Kleinrock conducted early research in queueing theory[4] which proved important in packet switching, and published a book in the related field of digital message switching (without the packets) in 1961; he also later played a leading role in building and management of the world's first packet-switched network, the ARPANET.

The concept of switching small blocks of data was first explored independently by Paul Baran at the RAND Corporation in the U. S. and Donald Davies at the National Physical Laboratory (NPL) in the UK in the early-to-mid-1960s.

Baran developed the concept of distributed adaptive message block switching during his research at the RAND Corporation for the U. S. Air Force into communications networks, that could survive nuclear wars, first presented to the Air Force in the summer of 1961 as briefing B-265, later published as RAND report P-2626 in 1962, and finally in report RM 3420 in 1964. Report P-2626 described a general architecture for a large-scale, distributed, survivable communications network. The work focuses on three key ideas: use of a decentralized network with multiple paths between any two points, dividing user messages into message blocks, later called packets, and delivery of these messages by store and forward switching.

Baran's work was known to Robert Taylor and J. C. R. Licklider at the Information Processing Technology Office, who advocated wide area networks, and it influenced Lawrence Roberts to adopt the technology in the development of the ARPANET.

Starting in 1965, Donald Davies at the National Physical Laboratory, UK, independently developed the same message routing methodology as developed by Baran. He called it packet switching, a more accessible name than Baran's, and proposed to build a nationwide network in the UK. He gave a talk on the proposal in 1966, after which a person from the Ministry of Defence (MoD) told him about Baran's work. A member of Davies' team (Roger Scantlebury) met Lawrence Roberts at the 1967 ACM Symposium on Operating System Principles and suggested it for use in the ARPANET.

Davies had chosen some of the same parameters for his original network design as did Baran, such as a packet size of 1024 bits. In 1966, Davies proposed that a network should be built at the laboratory to serve the needs of NPL and prove the feasibility of packet switching. The NPL Data Communications Network entered service in 1970.

The first computer network and packet switching network deployed for computer resource sharing was the Octopus Network at the Lawrence Livermore National Laboratory that began connecting four Control Data 6600 computers to several shared storage devices (including an IBM 2321 Data Cell in 1968 and an IBM Photostore in 1970) and to several hundred Teletype Model 33 ASR terminals for time sharing use starting in 1968.

In 1973, Vint Cerf and Bob Kahn wrote the specifications for Transmission Control Protocol (TCP), an internetworking protocol for sharing resources using packet-switching among the nodes.

3. Connectionless and Connection-Oriented Modes

Packet switching may be classified into connectionless packet switching, also known as datagram switching, and connection-oriented packet switching, also known as virtual circuit switching.

Examples of connectionless protocols are Ethernet, Internet Protocol (IP), and the User Datagram Protocol (UDP). Connection-oriented protocols include X. 25, Frame Relay, Multiprotocol Label Switching (MPLS), and the Transmission Control Protocol (TCP).

In connectionless mode each packet includes complete addressing information. The packets are routed individually, sometimes resulting in different paths and out-of-order delivery. Each packet is labeled with a destination address, source address, and port numbers. It may also be labeled with the sequence number of the packet. This precludes the need for a dedicated path to help the packet find its way to its destination, but means that much more information is needed in the packet header, which is therefore larger, and this information needs to be looked up in power-hungry content-addressable memory. Each packet is dispatched and may go via different routes; potentially, the system has to do as much work for every packet as the connection-oriented system has to do in connection set-up, but with less information as to the application's requirements. At the destination, the original message/data is reassembled in the correct order, based on the packet sequence number. Thus a virtual connection, also known as a virtual circuit or byte stream is provided to the end-user by a transport layer protocol, although intermediate network nodes only provides a connectionless network layer service.

Connection-oriented transmission requires a setup phase in each involved node before any packet is transferred to establish the parameters of communication. The packets include a connection identifier rather than address information and are negotiated between endpoints so that they are delivered in order and with error checking. Address information is only transferred to each node during the connection set-up phase, when the route to the destination is discovered and an entry is added to the switching table in each network node through which the connection passes. The signaling protocols used allow the application to specify its requirements and discover link parameters. Acceptable values for service parameters may be negotiated. Routing a packet requires the node to look up the connection id in a table. The packet header can be small, as it only needs to contain this code and any information, such as length, timestamp, or sequence number, which is different for different packets.

New Words

packet	[ˈpækɪt]	*n.* 数据包，信息包；小包，小捆　*v.* 包装，打包
robustness	[rəʊˈbʌstnɪs]	*n.* ［自］鲁棒性；［计］稳健性；健壮性
header	[ˈhedə]	*n.* 头球；页眉；数据头；收割台
payload	[ˈpeɪləʊd]	*n.* 有效负荷；收费载重，酬载
resonance	[ˈrez(ə)nəns]	*n.* ［力］共振；共鸣；反响
implementer	[ˈɪmplɪmentə]	*n.* 实施者；制订人；实现器；实作器
traverse	[ˈtrævəs; trəˈvɜːs]	*v.* 穿过；反对；详细研究；在……来回移动
billable	[ˈbɪləbl]	*adj.* 可收取费用的，可计费的
diminish	[dɪˈmɪnɪʃ]	*v.* 使减少，使变小；减少，缩小，变小
decentralize	[diːˈsentrəˈlaɪz]	*n.* 分散　*v.* 分散集权；疏散
methodology	[meθəˈdɒlədʒi]	*n.* 方法学，方法论
datagram	[ˈdeɪtəgræm]	*n.* 数据报
preclude	[prɪˈkluːd]	*v.* 排除；妨碍；阻止
identifier	[aɪˈdentɪfaɪə]	*n.* 标识符，认同者；检验人，鉴定人

Notes

[1] Paul Baran（保罗·巴兰）是波兰裔美国工程师。20 世纪 60 年代初，巴兰在兰德（RAND）公司担任技术工程师。兰德组建于第二次世界大战之后，并扮演着美国军方智囊团的角色。在兰德工作期间，巴兰开发了一套新型通信系统。该系统的主要特色是：如果部分系统被核武器摧毁，整个通信系统仍能够保持运行。

[2] RAND Corporation（兰德公司）是美国最重要的以军事为主的综合性战略研究机构。它先以研究军事尖端科学技术和重大军事战略而著称于世，继而又扩展到内外政策各方面，逐渐发展成为一个研究政治、军事、经济科技、社会等各方面的综合性思想库，

被誉为现代智囊的“大脑集中营”“超级军事学院”，以及世界智囊团的开创者和代言人。它可以说是当今美国乃至世界最负盛名的决策咨询机构。

[3] Donald Davies 是英国计算机科学家。其主要贡献包括：参与了英国第一台计算机的研制；主持了英国第一个实验网的建设，开发了分组交换技术，使计算机能够彼此通信，并且使互联网成为可能。

[4] queuing theory（排队论），或称随机服务系统理论，是通过对服务对象到来及服务时间的统计研究，得出这些数量指标（等待时间、排队长度、忙期长短等）的统计规律，然后根据这些规律来改进服务系统的结构或重新组织被服务对象，使得服务系统既能满足服务对象的需要，又能使机构的费用最经济或某些指标最优。该理论广泛应用于计算机网络、生产、运输、库存等各项资源共享的随机服务系统。

Questions for Discussion

1. What is the feature of packet switching?
2. What does Baran report in 1962 focus on?
3. List some examples of connectionless protocols.

Unit 8

Text A

Information Security

Information security,[1] sometimes shortened to InfoSec, is the practice of preventing unauthorized access, use, disclosure, disruption, modification, inspection, recording or destruction of information. It is a general term that can be used regardless of the form the data may take (e. g. electronic, physical).

1. Overview

(1) IT Security

Sometimes referred to as computer security, information technology security (IT security) is information security applied to technology (most often some form of computer system). It is worthwhile to note that a computer does not necessarily mean a home desktop. A computer is any device with a processor and some memory. Such devices can range from non-networked standalone devices as simple as calculators, to networked mobile computing devices such as smartphones and tablet computers. IT security specialists are almost always found in any major enterprise/establishment due to the nature and value of the data within larger businesses. They are responsible for keeping all of the technology within the company secure from malicious cyber attacks that often attempt to breach into critical private information or gain control of the internal systems.

(2) Information Assurance

The act of providing trust of the information, that the confidentiality, integrity and availability (CIA) of the information are not violated, e. g. ensuring that data is not lost when critical issues arise. These issues include, but are not limited to: natural disasters, computer/server malfunction or physical theft. Since most information is stored on computers in our modern era, information assurance is typically dealt with by IT security specialists. A common method of providing information assurance is to have an off-site backup of the data in case one of the mentioned issues arise.

(3) Threats

Information security threats come in many different forms. Some of the most common threats today are software attacks, theft of intellectual property, identity theft, theft of equipment or information, sabotage, and information extortion. Most people have experienced software attacks of some sort. Viruses, worms, phishing attacks, and Trojan horses are a few common examples of software attacks. The theft of intellectual property has also been an extensive issue for many businesses in the IT field. Identity theft is the attempt to act as someone else usually to obtain that person's personal information or to take advantage of their access to vital information. Theft of equipment or information is becoming more prevalent today due to the fact that most devices today are mobile. Cell phones are prone to theft and have also become far more desirable as the amount of data capacity increases. Sabotage usually consists of the destruction of an organization's website in an attempt to cause loss of confidence on the part of its customers. Information extortion consists of theft of a company's property or information as an attempt to receive a payment in exchange for returning the information or property back to its owner, as with ransomware. There are many ways to help protect yourself from some of these attacks but one of the most functional precautions is user carefulness.

Governments, military, corporations, financial institutions, hospitals and private businesses amass a great deal of confidential information about their employees, customers, products, research and financial status. Most of this information is now collected, processed and stored on electronic computers and transmitted across networks to other computers.

Should confidential information about a business' customers or finances or new product line fall into the hands of a competitor or a black hat hacker, a business and its customers could suffer widespread, irreparable financial loss, as well as damage to the company's reputation. From a business perspective, information security must be balanced against cost; the Gordon-Loeb Model[2] provides a mathematical economic approach for addressing this concern.

For the individual, information security has a significant effect on privacy, which is viewed very differently in various cultures.

The field of information security has grown and evolved significantly in recent years. It offers many areas for specialization, including securing networks and alliedinfrastructure, securing applications and databases, security testing, information systems auditing, business continuity planning and digital forensics.

(4) Responses to Threats

Possible responses to a security threat or risk are:

- Reduce/mitigate—implement safeguards and countermeasures to eliminate vulnerabilities or block threats.
- Assign/transfer—place the cost of the threat onto another entity or organization such as purchasing insurance or outsourcing.
- Accept—evaluate if cost of countermeasure outweighs the possible cost of loss due to threat.
- Ignore/reject—not a valid or prudent due-care response.

2. Definitions

The definitions of InfoSec suggested in different sources are summarized below (adopted from).

1) "Preservation of confidentiality, integrity and availability of information. Note: In addition, other properties, such as authenticity, accountability, non-repudiation and reliability can also be involved." (ISO/IEC 27000: 2009)

2) "The protection of information and information systems from unauthorized access, use, disclosure, disruption, modification, or destruction in order to provide confidentiality, integrity, and availability." (CNSS, 2010)

3) "Ensures that only authorized users (confidentiality) have access to accurate and complete information (integrity) when required (availability)." (ISACA, 2008)

4) "Information security is the process of protecting the intellectual property of an organisation." (Pipkin, 2000)

5) "... information security is a risk management discipline, whose job is to manage the cost of information risk to the business." (McDermott and Geer, 2001)

6) "A well-informed sense of assurance that information risks and controls are in balance." (Anderson, J., 2003)

7) "Information security is the protection of information and minimizes the risk of exposing information to unauthorized parties." (Venter and Eloff, 2003)

8) "Information security is a multidisciplinary area of study and professional activity which is concerned with the development and implementation of security mechanisms of all available types (technical, organizational, human-oriented and legal) in order to keep information in all its locations (within and outside the organization's perimeter) and, consequently, information systems, where information is created, processed, stored, transmitted and destroyed, free from threats. Threats to information and information systems may be categorized and a corresponding security goal may be defined for each category of threats. A set of security goals, identified as a result of a threat analysis, should be revised periodically to ensure its adequacy and conformance with the evolving environment. The currently relevant set of security goals may include: confidentiality, integrity, availability, privacy, authenticity & trustworthiness, non-repudiation, accountability and auditability." (Cherdantseva and Hilton, 2013)

3. Basic Principles

(1) Key Concepts

The CIA triad of confidentiality, integrity, and availability is at the heart of information security. (The members of the classic InfoSec triad—confidentiality, integrity and availability—are interchangeably referred to in the literature as security attributes, properties, security goals, fundamental aspects, information criteria, critical information characteristics and basic building blocks.) There is continuous debate about extending this classic trio. Other principles such as accountability have sometimes been proposed for addition—it has been pointed out that issues such as non-repudiation do not fit well within the three core concepts.

In 1992 and revised in 2002, the OECD's *Guidelines for the Security of Information Systems and Networks* proposed the nine generally accepted principles: awareness, responsibility, response, ethics, democracy, risk assessment, security design and implementation, security management, and reassessment. Building upon those, in 2004 the NIST's *Engineering Principles for Information Technology Security* proposed 33 principles. From each of these derived guidelines and practices.

In 2002, Donn Parker proposed an alternative model for the classic CIA triad that he called the six atomic elements of information. The elements are confidentiality, possession, integrity, authenticity, availability, and utility. The merits of the Parkerian Hexad are a subject of debate amongst security professionals.

In 2011, The Open Group published the information security management standard O-ISM3. This standard proposed an operational definition of the key concepts of security, with elements called "security objectives", related to access control, availability, data quality, compliance and technical. This model is not currently widely adopted.

(2) Confidentiality

In information security, confidentiality "is the property, that information is not made available or disclosed to unauthorized individuals, entities, or processes" (Excerpt ISO 27000).

(3) Integrity

In information security, data integrity means maintaining and assuring the accuracy and completeness of data over its entire life-cycle. This means that data cannot be modified in an unauthorized or undetected manner. This is not the same thing as referential integrity in databases, although it can be viewed as a special case of consistency as understood in the classic ACID[3] model of transaction processing. Information security systems typically provide message integrity in addition to data confidentiality.

(4) Availability

For any information system to serve its purpose, the information must be available when it is needed. This means that the computing systems used to store and process the information, the security controls used to protect it, and the communication channels used to access it must be functioning correctly. High availability systems aim to remain available at all times, preventing service disruptions due to power outages, hardware failures, and system upgrades. Ensuring availability also involves preventing denial-of-service attacks, such as a flood of incoming messages to the target system essentially forcing it to shut down.

(5) Non-repudiation

In law, non-repudiation implies one's intention to fulfill their obligations to a contract. It also implies that one party of a transaction cannot deny having received a transaction nor can the other party deny having sent a transaction.

It is important to note that while technology such as cryptographic systems can assist in non-repudiation efforts, the concept is at its core a legal concept transcending the realm of technology. It is not, for instance, sufficient to show that the message matches a digital signature signed with the sender's private key, and thus only the sender could have sent the message and nobody else could have altered it in transit (data integrity). The alleged sender could in return demonstrate that the digital signature algorithm is vulnerable or flawed, or allege or prove that his signing key has been compromised. The fault for these violations may or may not lie with the sender himself, and such assertions may or may not relieve the sender of liability, but the assertion would invalidate the claim that the signature necessarily proves authenticity and integrity; and, therefore, the sender may repudiate the message (because authenticity and integrity are pre-requisites for non-repudiation).

New Words

unauthorized	[ˌʌnˈɔːθəraɪzd]	*adj.* 非法的；未被授权的；独断的
desktop	[ˈdesktɒp]	*n.* 桌面；台式机
cyber	[ˈsaɪbə]	*adj.* 网络的，计算机的
confidentiality	[ˌkɒnfɪ, denʃɪˈælɪti]	*n.* 机密，[计] 机密性
integrity	[ɪnˈtegrɪti]	*n.* [计算机] 保存完整；诚实
malfunction	[mælˈfʌŋ(k)ʃ(ə)n]	*v.* 发生故障；不起作用 *n.* 故障；失灵
off-site	[ɒfset]	*adj.* 界外的；装置外的
backup	[ˈbækʌp]	*adj.* 支持的；备份的 *v.* 做备份
extortion	[ɪkˈstɔːʃ(ə)n; ek-]	*n.* 勒索；[法] 敲诈；强夺
prevalent	[ˈprev(ə)l(ə)nt]	*adj.* 流行的；普遍的，广传的
ransomware	[ˈransəmˌweə (r)]	*n.* 勒索软件
audit	[ˈɔːdɪt]	*v./n.* 审计；检查
forensics	[fəˈrensɪks]	*n.* 取证
mitigate	[ˈmɪtɪgeɪt]	*v.* 使缓和，使减轻
countermeasure	[ˈkaʊntəmeʒə]	*n.* 对策，反措施，对抗（或报复）手段
outweigh	[aʊtˈweɪ]	*v.* 比……重（在重量上）；比……有价值
authenticity	[ɔːθenˈtɪsɪti]	*n.* 真实性，确实性；可靠性
accountability	[əˌkaʊntəˈbɪlɪti]	*n.* 有义务；有责任；可说明性
property	[ˈprɒpəti]	*n.* 性质，性能；财产；所有权
multidisciplinary	[mʌltɪdɪsəˈplɪnəri]	*adj.* 多学科的
adequacy	[ˈædɪkwəsi]	*n.* 足够；适当；妥善性
conformance	[kənˈfɔːm(ə)ns]	*n.* 一致性；顺应
trio	[ˈtriːəʊ]	*n.* 三件一套；三个一组；三重唱
triad	[ˈtraɪæd]	*n.* 三和音；三个一组；三价元素
compliance	[kəmˈplaɪəns]	*n.* 顺从，服从；承诺
entity	[ˈentɪti]	*n.* 实体；存在；本质
referential	[ˌrefəˈrenʃ(ə)l]	*adj.* 指示的；用作参考的
outage	[ˈaʊtɪdʒ]	*n.* 储运损耗；中断供应；运行中断
repudiation	[rɪˌpjuːdɪˈeɪʃn]	*n.* 否认，拒绝；抛弃，断绝关系
cryptographic	[ˌkrɪptəˈgræfɪk]	*adj.* 用密码写的，关于暗号的

realm [relm] n. 领域，范围；王国
assertion [əˈsɜːʃ(ə)n] n. 断言，主张，要求；认定
invalidate [ɪnˈvælɪdeɪt] v. 使无效；使无价值
pre-requisite [ˌpriːˈrekwɪzɪt] adj. 必需的；首要的 n. 必要的事；首要的事

Notes

[1] information security：信息安全是指信息系统（包括硬件、软件、数据、人、物理环境及其基础设施）受到保护，不受偶然的或者恶意的原因而遭到破坏、更改、泄露，系统连续可靠正常地运行，信息服务不中断，最终实现业务连续性。

[2] Gordon-Loeb Model：戈登-洛布模型（Gordon-Loeb Model）是分析最优信息安全投资水平的数理经济学模型。模型指出在一般情况下，一个公司需要用于保护信息安全的花费仅应当是预期损失（信息安全漏洞所造成的损失的预期值）的一小部分。更为具体地说，多于信息安全漏洞预期损失37%的信息安全（包括网络安全）投资，通常是不经济的。同时，戈登-洛布模型指出，针对一定水平的潜在损失，用于保护一个信息集合的最优投资水平并不总随着信息集脆弱性的增强而增加。

[3] ACID：是指数据库事务正确执行的四个基本要素的缩写。它包含原子性（atomicity）、一致性（consistency）、隔离性（isolation）、持久性（durability）。一个支持事务（transaction）的数据库，必须具有这四种特性，否则在事务过程（transaction processing）当中无法保证数据的正确性，交易过程极可能达不到交易方的要求。

Questions for Discussion

1. What does information security refer to?
2. What is IT security?
3. What are the most common threats to information security?

Text B

5 Information Security Trends That Will Dominate 2016

Every year, it seems, the threats posed by cybercriminals evolve into new and more dangerous forms while security organizations struggle to keep up.

As 2015 draws to a close, we can expect the size, severity and complexity of cyber threats to continue increasing in 2016, says Steve Durbin, managing director for the Information Security Forum (ISF), a nonprofit association that assesses security and risk management issues on behalf of its members.

"For me, 2016 is probably the year of cyber risk," Durbin says. "I say that because increasingly I think we are seeing a raised level awareness about the fact that operating in cyber brings about its own peculiarities."

"As we move into 2016, attacks will continue to become more innovative and sophisticated,"

Durbin says. "Unfortunately, while organizations are developing new security mechanisms, cybercriminals are cultivating new techniques to evade them. In the drive to become more cyber resilient, organizations need to extend their risk management focus from pure information confidentiality, integrity and availability to include risks such as those to reputation and customer channels, and recognize the unintended consequences from activity in cyberspace. By preparing for the unknown, organizations will have the flexibility to withstand unexpected, high impact security events."

1. The Unintended Consequences of State Intervention

Conflicting official involvement in cyberspace in 2016 will create the threat of collateral damage and have unforeseen implications and consequences for all organizations that rely on it, Durbin says, noting that varying regulation and legislation will restrict activities whether or not an organization is the intended target. He warns that even organizations not implicated in wrongdoing will suffer collateral damage as authorities police their corner of the Internet.

"We've seen the European Court of Justice kicking out *Safe Harbor*[1]," Durbin says. "We're seeing increasing calls for backdoors from governments, while certain technology vendors are saying, 'Good luck, because we encrypt everything end-to-end and we have no knowledge of what this data is.' In a world where terrorism is becoming more the norm, there is a cyber-physical link here. How do we legislate in the face of that?"

Moving forward, Durbin says, organizations will have to understand what governments are able to ask for and be open about that with partners. "Legislators will always be playing catch up, and I think legislators themselves need to raise their game," Durbin says. "They'll always be talking about yesterday, and cyber is talking about tomorrow."

2. Big Data[2] Will Lead to Big Problems

Organizations are increasingly embedding big data in their operations and decision-making process. But it's essential to recognize that there is a human element to data analytics. Organizations that fail to respect that human element will put themselves at risk by overvaluing big data output, Durbin says, noting that poor integrity of the information sets could result in analyses that lead to poor business decisions, missed opportunities, brand damage and lost profits.

"There is this huge temptation that, of course, if you've accessed, it must be right," Durbin says. "This issue of data integrity, for me, is a big one. Sure, data is the lifeblood of an organization, but do we really know whether it's 'A-neg' or 'O-neg'?"

"There's this massive amount of information out there," he adds. "One of the things that scares me to death is not necessarily people stealing that information but actually manipulating it in ways that you're never going to see."

For instance, he notes that organizations have outsourced[3] code creation for years. "We don't know for certain that there aren't back doors in that code," he says. "In fact, there probably are. You're going to need to be much more skeptical about this: Question assumptions and make sure the information is actually what it says it is." And, of course, it's not simply the integrity of code you need to worry about. You need to understand the provenance of all your data. "If it's our information, we understand the provenance, that's fine," he says. "As soon as you start sharing it, you open yourself up. You need to know how the information is being used, who it's being shared with, who's adding to it and how it's being manipulated."

3. Mobile Applications and the IoT

"Smart phones and other mobile devices are creating a prime target for malicious actors in the Internet of Things (IoT)," Durbin says. The rapid uptake of bring-your-own-device (BYOD), and the introduction of wearable technologies to the workplace, will increase an already high demand for mobile apps for work and home in the coming year. To meet this increased demand, developers working under intense pressure and on razor-thin profit margins will sacrifice security and thorough testing in favor of speed of delivery and low cost, resulting in poor quality products more easily hijacked by criminals or hacktivists. "Don't confuse this with phones," Durbin says. "Mobility is more than that. The smart phone is just one component of mobility." He notes that there are an increasing number of workers just like him that are constantly mobile. "We don't have offices, as such," he says. "The last time I checked in it was a hotel. Today it's somebody else's office environment. How do I really know that it is 'Steve' coming in to this particular system? I might know that it's Steve's device, or what I believe to be Steve's device, but how do I know that it's Steve on the other end of that device?"

"Organizations should be prepared to embrace the increasingly complex IoT and understand what it means for them," Durbin says. Chief Information Security Officers (CISOs)[4] should be proactive in preparing the organization for the inevitable unknown situation by ensuring that apps developed in-house follow the testing steps in a recognized systems development lifecycle approach. They should also be managing user devices in line with existing asset management policies and processes, incorporating user devices into existing standards for access management and promoting education and awareness of BYOD risk in innovative ways.

4. Cybercrime Causes the Perfect Threat Storm

"Cybercrime topped the list of threats in 2015, and it's not going away in 2016," Durbin says. Cybercrime, along with an increase in hacktivism, the surge in cost of compliance to deal with the uptick in regulatory requirements and the relentless advances in technology against a backdrop of under investment in security departments, can all combine to cause the perfect threat storm. Organizations that adopt a risk management approach to identify what the business relies on most will be well placed to quantify the business case to invest in resilience.

Cyberspace is an increasingly attractive hunting ground for criminals, activists and terrorists motivated to make money, cause disruption or even bring down corporations and governments through online attacks. Organizations must be prepared for the unpredictable so they have the resilience to withstand unforeseen, high impact events.

"I see an increasing maturity and development of the cybercrime gangs," Durbin says. "They're incredibly sophisticated and well-coordinated. We're seeing an increase in crime as a service. This increasing sophistication is going to cause real challenges for organizations. We're really moving into an area where you can't predict how a cybercriminal is going to come after you. From an organizational standpoint, how do you defend against that?

Part of the problem is that many organizations are still focusing on defending the perimeter in an era when insiders—whether malicious or simply ignorant of proper security practices—make that perimeter increasingly permeable.

"We have viewed cybercrime rightly or wrongly from the perspective of it being an external attack, so we attempt to throw a security blanket over the perimeter if you will," Durbin says.

"There is a threat within. That takes us to a very uncomfortable place from an organizational standpoint."

The fact of the matter is that organizations won't be able to come to grips with cybercriminals unless they adopt a more forward-looking approach.

"A few weeks ago, I was speaking to a CISO of a major company with nine years on the job," Durbin says. "He told me that with big data analytics, he now has almost complete visibility across the entire organization. After nine years. The cybercriminals have had that capability for ages. Our approach is continually reactive as opposed to proactive."

"Cybercriminals don't work that way—based on history," he adds. "They're always trying to come up with a new way. I think we're still not that great at playing a defensive game. We need to really raise it to the same level. We're never going to be as imaginative. There's still this view inside the company that we haven't been broken into already, why are we spending all this money?"

5. Skills Aap Becomes an Abyss for Information Security

The information security professionals are maturing just as the increasing sophistication of cyber-attack capabilities demand more increasingly scarce information security professionals. While cybercriminals and hacktivists are increasing in numbers and deepening their skillsets, the "good guys" are struggling to keep pace, Durbin says. CISOs need to build sustainable recruiting practices and develop and retain existing talent to improve their organization's cyber resilience.

"The problem is going to grow worse in 2016 as hyper connectivity increases," Durbin says. CISOs will have to become more aggressive about getting the skill sets the organization needs.

"In 2016, I think we're going to become very much more aware that perhaps we don't have the right people in our security departments," he says. "We know that we've got some good technical guys who can fix firewalls and that sort of thing. But the right sort of people can make the case for cybersecurity being linked to business challenges and business developments. That's going to be a significant weakness. Boards are coming to the realization that cyber is the way they do business. We still don't have the joined up linkage between the business and the security practice."

In some cases, it's going to become apparent that organizations simply don't have the right CISO in place. Other organizations will have to ask themselves if security itself is sitting in the right place within the organization.

"You can't avoid every serious incident, and while many businesses are good at incident management, few have an established, organized approach for evaluating what went wrong," Durbin says. "As a result, they are incurring unnecessary costs and accepting inappropriate risks. Organizations of all sizes need to take stock now in order to ensure they are fully prepared and engaged to deal with these emerging security challenges. By adopting a realistic, broad-based, collaborative approach to cyber security and resilience, government departments, regulators, senior business managers and information security professionals will better understand the true nature of cyber threats and how to respond quickly and appropriately."

This article was written by Thor Olavsrud from CIO.

New Words

severity	[sɪ'verɪti]	n. 严重；严格；猛烈
peculiarity	[pɪˌkjuːlɪ'ærɪti]	n. 特性；特质；怪癖；奇特
resilient	[rɪ'zɪlɪənt]	adj. 弹回的，有弹力的；能复原的，有复原力的
collateral	[kə'læt(ə)r(ə)l]	n. 抵押品 adj. 并行的；旁系的；附属的
vendor	['vendə; 'vendɔː]	n. 卖主；小贩；供应商；[贸易]自动售货机
embed	[ɪm'bed; em-]	v. 栽种；使嵌入，使插入；使深留脑中
analytics	[ænə'lɪtɪks]	n. [化学][数]分析学；解析学
lifeblood	['laɪfblʌd]	n. 生机的根源；生命必需的血液；命脉
outsource	[aʊt'sɔːs]	v. 把……外包；外包
skeptical	['skeptɪkəl]	adj. 怀疑的；怀疑论的，不可知论的
innovative	['ɪnəvətɪv]	adj. 革新的，创新的；新颖的；有创新精神的
uptick	['ʌptɪk]	n. 报升（股票成交价格比上一个交易的为高）
regulatory	['regjʊlətəri]	adj. 管理的；控制的；调整的
quantify	['kwɒntɪfaɪ]	v. 量化，为……定量，确定数量；定量
resilience	[rɪ'zɪlɪəns]	n. 恢复力；弹力；顺应力
proactive	[prəʊ'æktɪv]	adj. 有前瞻性的，先行一步的；积极主动的
incur	[ɪn'kɜː]	v. 招致，引发；蒙受

Notes

[1] Safe Harbor（安全港协议）是2000年12月美国商业部与欧盟建立的协议，它用于调整美国企业出口以及处理欧洲公民的个人数据。美国与欧盟签署的个人信息跨国流通安全港协议暗示，在商业利益的驱动下，以美国为代表的行业自律和市场调节机制为主的松散立法体制向以欧盟为代表的统一立法体制暂时做出了让步。在该协议中，美欧就其协议基本原则和取得，执行以及制裁机制做出了一系列具体规定。

[2] Big Data（大数据），或称巨量资料，是指所涉及的资料量规模巨大到无法通过目前主流软件工具，在合理时间内达到撷取、管理、处理，并整理成为帮助企业提升经营决策的资讯。大数据谈的不仅仅是数据量，其实包含了数据量（Volume）、时效性

(Velocity)、多样性（Variety）和真实性（Veracity）。

[3] outsource：是指外购（是指从外国供应商等处获得货物或服务）或外包（工程）。

[4] CISO是首席信息安全官，主要负责对机构内的信息安全进行评估、管理和实现。CISO也称为IT安全主管，安全行政官或类似的称呼。CISO通常直接向CIO汇报，但在较大的机构里，两者之间常常还有一个或多个管理层。然而，CISO给CIO提出的建议并不是辅助的、不重要的问题，这些建议即使不比其他技术和信息相关的决策更重要，至少也是同等重要的。

Questions for Discussion

1. What should organizations do in the drive to become more cyber resilient?
2. What should Chief Information Security Officers (CISOs) do when preparing to embrace the increasingly complex IoT?
3. How can government departments, regulators, senior business managers and information security professionals better understand the true nature of cyber threats?

Unit 9

Text A

Multiplexing and Multiple Access

1. Multiplexing

Because of the installation cost of a communications channel, such as a microwave link or a coaxial cable link, it is desirable to share the channel among multiple users. Provided that the channel's data capacity exceeds that required to support a single user, the channel may be shared through the use of multiplexing methods. Multiplexing is the sharing of a communications channel through local combining of signals at a common point. Two types of multiplexing are commonly employed: frequency-division multiplexing and time-division multiplexing.

(1) Frequency-Division Multiplexing

In frequency-division multiplexing (FDM)[1], the available bandwidth of a communications channel is shared among multiple users by frequency translating, or modulating, each of the individual users onto a different carrier frequency. Assuming sufficient frequency separation of the carrier frequencies that the modulated signals do not overlap, recovery of each of the FDM signals is possible at the receiving end. In order to prevent overlap of the signals and to simplify filtering, each of the modulated signals is separated by a guard band, which consists of an unused portion of the available frequency spectrum. Each user is assigned a given frequency band for all time.

While each user's information signal may be either analog or digital, the combined FDM signal is inherently an analog waveform. Therefore, an FDM signal must be transmitted over an analog channel. Examples of FDM are found in some of the old long-distance telephone transmission systems, including the American N-and L-carrier coaxial cable systems and analog point-to-point microwave systems. In the L-carrier system a hierarchical combining structure is employed in which 12 voiceband signals are frequency-division multiplexed to form a group signal in the frequency range of 60 to 108 kilohertz. Five group signals are multiplexed to form a supergroup signal in the frequency range of 312 to 552 kilohertz, corresponding to 60 voiceband signals, and 10 supergroup signals are multiplexed to form a master group signal. In the L1 carrier system, deployed in the 1940s, the master group was transmitted directly over coaxial cable. For microwave systems, it was frequency modulated onto a microwave carrier frequency for point-to-point transmission. In the L4 system, developed in the 1960s, six master groups were combined to form a jumbo group signal of 3, 600 voiceband signals.

(2) Time-Division Multiplexing

Multiplexing also may be conducted through the interleaving of time segments from different signals onto a single transmission path—a process known as time-division multiplexing (TDM)[2]. Time-division multiplexing of multiple signals is possible only when the available data rate of the channel exceeds the data rate of the total number of users. While TDM may be applied to either digital or analog signals, in practice it is applied almost always to digital signals. The resulting composite signal is thus also a digital signal.

In a representative TDM system, data from multiple users are presented to a time-division multiplexer. A scanning switch then selects data from each of the users in sequence to form a composite TDM signal consisting of the interleaved data signals. Each user's data path is assumed to be time-aligned or synchronized to each of the other users' data paths and to the scanning mechanism. If only one bit were selected from each of the data sources, then the scanning mechanism would select the value of the arriving bit from each of the multiple data sources. In practice, however, the scanning mechanism usually selects a slot of data consisting of multiple bits of each user's data; the scanner switch is then advanced to the next user to select another slot, and so on. Each user is assigned a given time slot for all time.

Most modern telecommunications systems employ some form of TDM for transmission over long-distance routes. The multiplexed signal may be sent directly over cable systems, or it may be modulated onto a carrier signal for transmission via radio wave. Examples of such systems include the North American T carriers as well as digital point-to-point microwave systems. In T1 systems, introduced in 1962, 24 voiceband signals (or the digital equivalent) are time-division multiplexed together. The voiceband signal is a 64-kilobit-per-second data stream consisting of 8-bit symbols transmitted at a rate of 8, 000 symbols per second. The TDM process interleaves 24 8-bit time slots together, along with a single frame-synchronization bit, to form a 193-bit frame. The 193-bit frames are formed at the rate of 8, 000 frames per second, resulting in an overall data rate of 1. 544 megabits per second. For transmission over more recent T-carrier systems, T1 signals are often further multiplexed to form higher-data-rate signals—again using a hierarchical scheme.

2. Multiple Access

Multiplexing is defined as the sharing of a communications channel through local combining at a common point. In many cases, however, the communications channel must be efficiently shared among many users that are geographically distributed and that sporadically attempt to communicate at random points in time. Three schemes have been devised for efficient sharing of a single channel under these conditions; they are called frequency-division multiple access (FDMA), time-division multiple access (TDMA), and code-division multiple access (CDMA)[3]. These techniques can be used alone or together in telephone systems, and they are well illustrated by the most advanced mobile cellular systems.

(1) Frequency-Division Multiple Access

In FDMA the goal is to divide the frequency spectrum into slots and then to separate the signals of different users by placing them in separate frequency slots. The difficulty is that the frequency spectrum is limited and that there are typically many more potential communicators than there are available frequency slots. In order to make efficient use of the communications channel, a system must be devised for managing the available slots. In the advanced mobile phone system (AMPS), the cellular system employed in the United States, different callers use separate frequency slots via FDMA. When one telephone call is completed, a network-managing computer at the cellular base station reassigns the released frequency slot to a new caller. A key goal of the AMPS system is to reuse frequency slots whenever possible in order to accommodate as many callers as possible. Locally within a cell, frequency slots can be reused when corresponding calls are terminated. In addition, frequency slots can be used simultaneously by multiple callers located in separate cells. The cells must be far enough apart geographically that the radio signals from one cell are sufficiently attenuated at the location of the other cell using the same frequency slot.

(2) Time-Division Multiple Access

In TDMA the goal is to divide time into slots and separate the signals of different users by placing the signals in separate time slots. The difficulty is that requests to use a single communications channel occur randomly, so that on occasion the number of requests for time slots is greater than the number of available slots. In this case information must be buffered, or stored in memory, until time slots become available for transmitting the data. The buffering introduces delay into the system. In the IS54 cellular system, three digital signals are interleaved using TDMA and then transmitted in a 30-kilohertz frequency slot that would be occupied by one analog signal in AMPS[4]. Buffering digital signals and interleaving them in time causes some extra delay, but the delay is so brief that it is not ordinarily noticed during a call. The IS54 system uses aspects of both TDMA and FDMA.

(3) Code-Division Multiple Access

In CDMA, signals are sent at the same time in the same frequency band. Signals are either selected or rejected at the receiver by recognition of a user-specific signature waveform, which is constructed from an assigned spreading code. The IS95 cellular system employs the CDMA technique. In IS95 an analog speech signal that is to be sent to a cell site is first quantized and then organized into one of a number of digital frame structures. In one frame structure, a frame of 20 milliseconds' duration consists of 192 bits. Of these 192 bits, 172 represent the speech signal itself, 12 form a cyclic redundancy check that can be used for error detection, and 8 form an encoder "tail" that allows the decoder to work properly. These bits are formed into an encoded data stream. After interleaving of the encoded data stream, bits are organized into groups of six. Each group of six bits indicates which of 64 possible waveforms to transmit. Each of the waveforms to be transmitted has a particular pattern of alternating polarities and occupies a certain portion of the radio-frequency spectrum. Before one of the waveforms is transmitted, however, it is multiplied by a code sequence of polarities that alternate at a rate of 1.2288 megahertz, spreading the bandwidth occupied by the signal and causing it to occupy (after filtering at the transmitter) about 1.23 megahertz of the radio-frequency spectrum. At the cell site one user can be selected from multiple users of the same 1.23-megahertz bandwidth by its assigned code sequence.

CDMA is sometimes referred to as spread-spectrum multiple access (SSMA), because the process of multiplying the signal by the code sequence causes the power of the transmitted signal to be spread over a larger bandwidth. Frequency management, a necessary feature of FDMA, is eliminated in CDMA. When another user wishes to use the communications channel, it is assigned a code and immediately transmits instead of being stored until a frequency slot opens.

New Words

access	[ˈækses]	*n.* 进入；使用权；通路
desirable	[dɪˈzaɪərəb(ə)l]	*adj.* 令人满意的；值得要的
modulate	[ˈmɒdjʊleɪt]	*v.* 调节；(信号)调制；调整
overlap	[əʊvəˈlæp]	*n.* 重叠；重复 *v.* 部分重叠；部分的同时发生
simplify	[ˈsɪmplɪfaɪ]	*v.* 简化；使单纯；使简易
assign	[əˈsaɪn]	*v.* 分配；指派；[计][数] 赋值
transmit	[trænzˈmɪt]	*v.* 传输；传播；发射；传达；遗传
hierarchical	[haɪəˈrɑːkɪk(ə)l]	*adj.* 分层的；分级的；等级体系的
kilohertz	[ˈkɪləhɜːts]	*n.* [物] 千赫
jumbo	[ˈdʒʌmbəʊ]	*adj.* 巨大的；特大的 *n.* 庞然大物
segment	[ˈsegm(ə)nt]	*v.* 分割 *n.* 段；部分
composite	[ˈkɒmpəzɪt]	*n.* 合成物 *adj.* 合成的 *v.* 使合成；使混合
interleave	[ɪntəˈliːv]	*v.* [计] 交错，交叉存取
align	[əˈlaɪn]	*v.* 使结盟；使成一行；匹配
mechanism	[ˈmek(ə)nɪz(ə)m]	*n.* 机制；原理，途径；进程；机械装置；技巧
slot	[slɒt]	*n.* 位置；狭槽；硬币投币口 *v.* 跟踪；开槽于
sporadically	[spəˈrædɪkəli]	*adv.* 零星地；偶发地
simultaneously	[ˌsɪmlˈteɪnɪəsli]	*adv.* 同时地
attenuate	[əˈtenjʊeɪt]	*adj.* 减弱的；细小的 *v.* 使减弱；使纤细
buffer	[ˈbʌfə]	*n.* [计] 缓冲区；缓冲器 *v.* 缓冲
duration	[djʊˈreɪʃ(ə)n]	*n.* 持续，持续的时间，期间
polarity	[pə(ʊ)ˈlærɪti]	*n.* [物] 极性；两极；对立

Notes

[1] frequency-division multiplexing (FDM) 意为“频分多路复用”，是指用不同频率传送各路消息，以实现多路通信。这种方法也称为频率复用。

[2] TDM 是指时分复用模式。时分复用是指一种通过不同信道或时隙中的交叉位脉冲，同

时在同一个通信媒体上传输多个数字化数据、语音和视频信号等的技术。

[3] CDMA 是指一种扩频多址数字式通信技术，通过独特的代码序列建立信道，可用于二代和三代无线通信中的任何一种协议。CDMA 是一种多路方式，多路信号只占用一条信道，可以极大提高带宽使用率，应用于 800MHz 和 1.9GHz 的超高频（UHF）移动电话系统。

[4] AMPS 是第一代蜂窝技术，使用单独的频带，或者说“信道”，为每次对话服务。它因此需要相当的带宽来支持一个大数量的用户群体。在通用术语中，AMPS 常常被当作更早的“0G”改进型移动通信服务，只不过 AMPS 使用更多的计算功率来选择频谱，切换到 PSTN 线路的通话，以及处理登记和呼叫建立。

Questions for Discussion

1. In this passage, what does multiplexing mean?
2. Which two kinds of multiplexing are frequently used in order for multiple users to share the communications channel?
3. What are the three schemes devised for efficient sharing of a single channel?

Text B

Orthogonal Frequency-Division Multiplexing

Orthogonal frequency-division multiplexing (OFDM)[1] is a method of encoding digital data on multiple carrier frequencies. OFDM has developed into a popular scheme for wideband digital communication, used in applications such as digital television and audio broadcasting, DSL Internet access, wireless networks, powerline networks, and 4G mobile communications.

OFDM is a frequency-division multiplexing (FDM) scheme used as a digital multi-carrier modulation method. A large number of closely spaced orthogonal sub-carrier signals are used to carry data on several parallel data streams or channels. Each sub-carrier is modulated with a conventional modulation scheme (such as quadrature amplitude modulation or phase-shift keying) at a low symbol rate, maintaining total data rates similar to conventional single-carrier modulation schemes in the same bandwidth.

The primary advantage of OFDM over single-carrier schemes is its ability to cope with severe channel conditions (for example, attenuation of high frequencies in a long copper wire, narrowband interference and frequency-selective fading due to multipath) without complex equalization filters. Channel equalization is simplified because OFDM may be viewed as using many slowly modulated narrowband signals rather than one rapidly modulated wideband signal. The low symbol rate makes the use of a guard interval between symbols affordable, making it possible to eliminate intersymbol interference (ISI)[2] and utilize echoes and time-spreading (on analogue TV these are visible as ghosting and blurring, respectively) to achieve a diversity gain, i. e. a signal-to-noise ratio improvement. This mechanism also facilitates the design of single frequency networks (SFNs), where several adjacent transmitters send the same signal simultaneously at the same frequency, as the signals from multiple distant transmitters may be combined constructively, rather than interfering as would typically occur in a traditional single-carrier system.

1. Example of Applications

The following list is a summary of existing OFDM-based standards and products.

Wired

- ADSL and VDSL[3] broadband access via POTS copper wiring;
- DVB-C2, an enhanced version of the DVB-C digital cable TV standard;
- Power line communication (PLC);
- ITU-T G. hn, a standard which provides high-speed local area networking of existing home wiring (power lines, phone lines and coaxial cables);
- TrailBlazer telephone line modems;
- Multimedia over Coax Alliance (MoCA) home networking;
- DOCSIS[4] 3. 1 Broadband delivery.

Wireless

- The wireless LAN (WLAN) radio interfaces IEEE 802. 11a, g, n, ac and HIPERLAN/2;
- The digital radio systems DAB/EUREKA 147, DAB +, Digital Radio Mondiale, HD Radio, T-DMB and ISDB-TSB;
- The terrestrial digital TV systems DVB-T and ISDB-T;
- The terrestrial mobile TV systems DVB-H, T-DMB, ISDB-T and MediaFLO forward link;
- The wireless personal area network (PAN) ultra-wideband (UWB) IEEE 802. 15. 3a implementation suggested by WiMedia Alliance.

The OFDM based multiple access technology OFDMA is also used in several 4G and pre-4G cellular networks and mobile broadband standards:

- The mobility mode of the wireless MAN/broadband wireless access (BWA) standard IEEE 802. 16e (or Mobile-WiMAX);
- The mobile broadband wireless access (MBWA) standard IEEE 802. 20;
- The downlink of the 3GPP Long Term Evolution (LTE) fourth generation mobile broadband standard. The radio interface was formerly named High Speed OFDM Packet Access (HSOPA), now named Evolved UMTS Terrestrial Radio Access (E-UTRA).

2. Key Features

Summary of advantages

- High spectral efficiency as compared to other double sideband modulation schemes, spread spectrum, etc. ;
- Can easily adapt to severe channel conditions without complex time-domain equalization;
- Robust against narrow-band co-channel interference;
- Robust against intersymbol interference (ISI) and fading caused by multipath propagation;

- Efficient implementation using fast Fourier transform (FFT);
- Low sensitivity to time synchronization errors;
- Tuned sub-channel receiver filters are not required (unlike conventional FDM);
- Facilitates single frequency networks (SFNs) (i. e. transmitter macrodiversity).

Summary of disadvantages

- Sensitive to Doppler shift;
- Sensitive to frequency synchronization problems;
- High peak-to-average-power ratio (PAPR), requiring linear transmitter circuitry, which suffers from poor power efficiency;
- Loss of efficiency caused by cyclic prefix/guard interval.

Orthogonality

Conceptually, OFDM is a specialized FDM, the additional constraint being that all carrier signals are orthogonal to one another.

In OFDM, the sub-carrier frequencies are chosen so that the sub-carriers are orthogonal to each other, meaning that cross-talk between the sub-channels is eliminated and inter-carrier guard bands are not required. This greatly simplifies the design of both the transmitter and the receiver; unlike conventional FDM, a separate filter for each sub-channel is not required.

The orthogonality requires that the sub-carrier spacing is $\Delta f = \frac{k}{T_U}$ Hertz, where T_U seconds is the useful symbol duration (the receiver-side window size), and k is a positive integer, typically equal to 1. Therefore, with N sub-carriers, the total passband bandwidth will be $B \approx N\Delta f$ (Hz).

The orthogonality also allows high spectral efficiency, with a total symbol rate near the Nyquist rate for the equivalent baseband signal (i. e. near half the Nyquist rate for the double-side band physical passband signal). Almost the whole available frequency band can be utilized. OFDM generally has a nearly 'white' spectrum, giving it benign electromagnetic interference properties with respect to other co-channel users.

A simple example: A useful symbol duration $T_U = 1$ ms would require a sub-carrier spacing of 1kHz (or an integer multiple of that) for orthogonality. N = 1, 000 sub-carriers would result in a total passband bandwidth of $N \cdot \Delta f = 1$ MHz. For this symbol time, the required bandwidth in theory according to Nyquist is BW $= R/2 = (N/T_U)/2 = 0.5$MHz (i. e., half of the achieved bandwidth required by our scheme). If a guard interval is applied, Nyquist bandwidth requirement would be even lower. The FFT would result in N = 1, 000 samples per symbol. If no guard interval was applied, this would result in a base band complex valued signal with a sample rate of 1 MHz, which would require a baseband bandwidth of 0. 5 MHz according to Nyquist. However, the passband RF signal is produced by multiplying the baseband signal with a carrier waveform (i. e., double-sideband quadrature amplitude-modulation) resulting in a passband bandwidth of 1 MHz. A single-side band (SSB) or vestigial sideband (VSB) modulation scheme would achieve almost half that bandwidth for the same symbol rate (i. e., twice as high spectral efficiency for the same symbol alphabet length). It is however more sensitive to multipath interference.

OFDM requires very accurate frequency synchronization between the receiver and the transmitter; with frequency deviation the sub-carriers will no longer be orthogonal, causing inter-carrier interference (ICI) (i. e. , cross-talk between the sub-carriers). Frequency offsets are typically caused by mismatched transmitter and receiver oscillators, or by Doppler shift due to movement. While Doppler shift alone may be compensated for by the receiver, the situation is worsened when combined with multipath, as reflections will appear at various frequency offsets, which is much harder to correct. This effect typically worsens as speed increases, and is an important factor limiting the use of OFDM in high-speed vehicles. In order to mitigate ICI[5] in such scenarios, one can shape each sub-carrier in order to minimize the interference resulting in a non-orthogonal subcarriers overlapping. For example, a low-complexity scheme referred to as WCP-OFDM (weighted cyclic prefix orthogonal frequency-division Multiplexing) consists of using short filters at the transmitter output in order to perform a potentially non-rectangular pulse shaping and a near perfect reconstruction using a single-tap per subcarrier equalization. Other ICI suppression techniques usually increase drastically the receiver complexity.

New Words

orthogonal	[ɔːˈθɒg(ə)n(ə)l]	*n.* 正交直线 *adj.* [数] 正交的；直角的
encode	[ɪnˈkəʊd]	*v.* （将文字材料）译成密码；编码
carrier	[ˈkærɪə]	*n.* 载波
attenuation	[əˌtenjʊˈeɪʃən]	*n.* [物] 衰减；变薄；稀释
blur	[blɜː]	*v.* 涂污；使……模糊不清；使暗淡；玷污
trailblazer	[ˈtreɪlbleɪzə]	*n.* 开拓者；开路的人；先驱
adapt	[əˈdæpt]	*v.* 使适应；改编；适应
fade	[feɪd]	*v.* 褪色；凋谢；逐渐消失
macrodiversity	[ˈmækrəʊ-daɪˈvɜːsəti]	*n.* 宏分集
cyclic	[ˈsaɪklɪk]	*adj.* 环的；循环的；周期的
constraint	[kənˈstreɪnt]	*n.* [数] 约束；局促，态度不自然；强制
vestigial	[veˈstɪdʒɪəl]	*adj.* 退化的；残余的；发育不全的
subcarrier	[ˈsʌbˈkæriə]	*n.* [电子] [通信] 副载波；辅助波
suppression	[səˈpreʃ(ə)n]	*n.* 抑制；镇压；[植] 压抑
drastically	[ˈdræstɪkəli]	*adv.* 彻底地；激烈地

Notes

[1] orthogonal frequency division multiplexing (OFDM)：即正交频分多路复用技术，它是MCM (multi carrier modulation)，多载波调制的一种。OFDM 技术由 MCM (multi-carrier modulation，多载波调制) 发展而来。OFDM 技术是多载波传输方案的实现方式之一，它的调制和解调是分别基于 IFFT 和 FFT 来实现的，是实现复杂度最低、应用最广的一种多载波传输方案。

[2] intersymbol interference (ISI)：码间串扰，数字信号经常使用波形编码来使脉冲序列具

有特定的频谱特性。当脉冲通过限带信道时，脉冲会在时间上延伸，每个符号的脉冲将延伸到相邻的符号的时间间隔内。这就是码间串扰，并会导致接收机在检测一个符号时发生错误的概率分布。

[3] VDSL是一种非对称DSL技术，全称very high speed digital subscriber line（超高速数字用户线路）。和ADSL技术一样，VDSL也使用双绞线进行语音和数据的传输。VDSL是在现有电话线上安装的，只需在用户侧安装一台VDSL modem。最重要的是，无须为宽带上网而重新布设或变动线路。VDSL技术采用频分复用原理，数据信号和电话音频信号使用不同的频段，互不干扰，上网的同时可以拨打或接听电话。

[4] DOCSIS（data over cable service interface specifications）有线电缆数据服务接口规范，是一个由有线电缆标准组织Cable Labs制定的国际标准。

[5] ICI是interface control information的简称，表示接口控制信息，它是一种特殊的数据结构，它的格式可以用ICI Editor进行定义。

Questions for Discussion

1. What is the primary advantage of OFDM over single-carrier schemes?
2. List at least four advantages of OFDM.
3. What is the additional constraint of OFDM?

Unit 10

Text A

Modulation and Demodulation

In many telecommunications systems, it is necessary to represent an information-bearing signal with a waveform that can pass accurately through a transmission medium. This assigning of a suitable waveform is accomplished by modulation, which is the process by which some characteristic of a carrier wave is varied in accordance with an information signal, or modulating wave. The modulated signal is then transmitted over a channel, after which the original information-bearing signal is recovered through a process of demodulation. Modulation is applied to information signals for a number of reasons, some of which are outlined below.

1) Many transmission channels are characterized by limited passbands—that is, they will pass only certain ranges of frequencies without seriously attenuating them (reducing their amplitude). Modulation methods must therefore be applied to the information signals in order to "frequency translate" the signals into the range of frequencies that are permitted by the channel. Examples of channels that exhibit passband characteristics include alternating-current-coupled coaxial cables, which pass signals only in the range of 60 kilohertz to several hundred megahertz, and fibre-optic cables, which pass light signals only within a given wavelength range without significant attenuation. In these instances, frequency translation is used to "fit" the information signal to the communications channel.

2) In many instances a communications channel is shared by multiple users. In order to prevent mutual interference, each user's information signal is modulated onto an assigned carrier of a specific frequency. When the frequency assignment and subsequent combining is done at a central point, the resulting combination is a frequency-division multiplexed signal. Frequently there is no central combining point, and the communications channel itself acts as a distributed combine. An example of the latter situation is the broadcast radio bands (from 540 kilohertz to 600 megahertz), which permit simultaneous transmission of multiple AM radio, FM radio, and television signals without mutual

interference as long as each signal is assigned to a different frequency band.

3) Even when the communications channel can support direct transmission of the information-bearing signal, there are often practical reasons why this is undesirable. A simple example is the transmission of a three-kilohertz (i. e., voiceband) signal via radio wave. In free space the wavelength of a three-kilohertz signal is 100 kilometres. Since an effective radio antenna is typically as large as half the wavelength of the signal, a three-kilohertz radio wave might require an antenna up to 50 kilometres in length. In this case translation of the voice frequency to a higher frequency would allow the use of a much smaller antenna.

1. Analog modulation[1]

As is noted in analog-to-digital conversion, voice signals, as well as audio and video signals, are inherently analog in form. In most modern systems these signals are digitized prior to transmission, but in some systems the analog signals are still transmitted directly without converting them to digital form. There are two commonly used methods of modulating analog signals. One technique, called amplitude modulation, varies the amplitude of a fixed-frequency carrier wave in proportion to the information signal. The other technique, called frequency modulation, varies the frequency of a fixed-amplitude carrier wave in proportion to the information signal.

2. Digital Modulation[2]

In order to transmit computer data and other digitized information over a communications channel, an analog carrier wave can be modulated to reflect the binary nature of the digital baseband signal. The parameters of the carrier that can be modified are the amplitude, the frequency, and the phase.

3. Amplitude-Shift Keying[3]

If amplitude is the only parameter of the carrier wave to be altered by the information signal, the modulating method is called amplitude-shift keying (ASK). ASK can be considered a digital version of analog amplitude modulation. In its simplest form, a burst of radio frequency is transmitted only when a binary 1 appears and is stopped when a 0 appears. In another variation, the 0 and 1 are represented in the modulated signal by a shift between two preselected amplitudes.

4. Frequency-Shift Keying[4]

If frequency is the parameter chosen to be a function of the information signal, the modulation method is called frequency-shift keying (FSK). In the simplest form of FSK signaling, digital data is transmitted using one of two frequencies, whereby one frequency is used to transmit a 1 and the other frequency to transmit a 0. Such a scheme was used in the Bell 103 voiceband modem, introduced in 1962, to transmit information at rates up to 300 bits per second over the public switched telephone network. In the Bell 103 modem, frequencies of 1, 080 +/- 100 hertz and 1, 750 +/- 100 hertz were used to send binary data in both directions.

5. Phase-Shift Keying[5]

When phase is the parameter altered by the information signal, the method is called phase-shift keying (PSK). In the simplest form of PSK a single radio frequency carrier is sent with a fixed phase to represent a 0 and with a 180° phase shift—that is, with the opposite polarity—to represent a 1. PSK was employed in the Bell 212 modem, which was introduced about 1980 to transmit information at rates up to 1, 200 bits per second over the public switched telephone network.

6. Advanced Methods

In addition to the elementary forms of digital modulation described above, there exist more advanced methods that result from a superposition of multiple modulating signals. An example of the latter form of modulation is quadrature amplitude modulation (QAM)[6]. QAM signals actually transmit two amplitude-modulated signals in phase quadrature (i. e., 90° apart), so that four or more bits are represented by each shift of the combined signal. Communications systems that employ QAM include digital cellular systems in the United States and Japan as well as most voiceband modems transmitting above 2, 400 bits per second.

A form of modulation that combines convolutional codes with QAM is known as trellis-coded modulation (TCM)[7]. Trellis-coded modulation forms an essential part of most of the modern voiceband modems operating at data rates of 9, 600 bits per second and above, including V. 32 and V. 34 modems.

New Words

demodulation	[diːˌmɒdjʊˈleɪʃən]	*n.* 检波；反调制；解调制
amplitude	[ˈæmplɪtjuːd]	*n.* 振幅；丰富，充足
megahertz	[ˈmegəhɜːts]	*n.* 兆赫
interference	[ɪntəˈfɪər(ə)ns]	*n.* 干扰，冲突；干涉
antenna	[ænˈtenə]	*n.* 天线
conversion	[kənˈvɜːʃ(ə)n]	*n.* 转换；变换
digitize	[ˈdɪdʒɪtaɪz]	*v.* ［计］数字化
binary	[ˈbaɪnəri]	*adj.* ［数］二进制的；二元的，二态的
baseband	[ˈbeɪsbænd]	*n.* 基带
preselect	[ˌpriːsɪˈlekt]	*v.* 预选
whereby	[weəˈbaɪ]	*adv.* 凭借；通过……；借以；与……一致
modem	[ˈməʊdem]	*n.* 调制解调器（等于 modulator-demodulator）
superposition	[ˌsjuːpəpəˈzɪʃən]	*n.* ［数］叠加，重合
quadrature	[ˈkwɒdrətʃə]	*n.* 正交；求积；弦
convolutional	[ˌkɒnvəˈluːʃ(ə)nəl]	*adj.* 卷积的；回旋的；脑回的
trellis	[ˈtrelɪs]	*n.* 格子；格子结构；框架

Notes

[1] analog modulation：一般是指调制信号和载波都是连续波的调制方式。它有调幅、调频和调相三种基本形式。

[2] digital modulation：一般是指调制信号是离散的，而载波是连续波的调制方式。它有四种基本形式：振幅键控、移频键控、移相键控和差分移相键控。

[3] amplitude shift keying（ASK）：即幅移键控，是指振幅键控方式。这种调制方式是根据信号的不同，调节正弦波的振幅。幅度键控可以通过乘法器和开关电路来实现。

[4] frequency-shift keying（FSK）：频移键控，是指以数字信号控制载波频率变化的调制方式。根据已调波的相位连续与否，频移键控分为两类：相位不连续的频移键控和相位连续的频移键控。频移键控是信息传输中使用得较早的一种调制方式，它的主要优点是：实现起来较容易，抗噪声与抗衰减的性能较好。

[5] Phase-shift keying（PSK）：相位偏移调制，又称移相键控，是一种利用相位差异的信号来传送资料的调制方式。这种调制方式因此而得名。该传送信号必须为正交信号，其基底更须为单位化信号。

[6] quadrature amplitude modulation（QAM）：正交振幅键控是一种将两种调幅信号（2ASK和2PSK）汇合到一个信道的方法，因此会双倍扩展有效带宽，正交调幅被用于脉冲调幅。

[7] trellis coded modulation（TCM）：网格编码调制，在传统的数字传输系统中，发送端和接收端的纠错与调制电路是两个独立的部分，而纠错编码会带来频带利用率的下降。为了提高频带的利用率，同时也希望在不增加信道传输带宽的前提下降低差错率，可以把编码和调制相结合统一进行设计，这就是所谓的网格编码调制。

Questions for Discussion

1. What are the major reasons for application of modulation to information signals?
2. What are the frequently employed methods of modulating analog signals?
3. Do you know any other advanced forms of digital modulation?

Text B

Pulse-Code Modulation

Pulse-code modulation (PCM) is a method used to digitally represent sampled analog signals. It is the standard form of digital audio in computers, compact discs, digital telephony and other digital audio applications. In a PCM stream, the amplitude of the analog signal is sampled regularly at uniform intervals, and each sample is quantized to the nearest value within a range of digital steps. Linear pulse-code modulation (LPCM) is a specific type of PCM where the quantization levels are linearly uniform. This is in contrast to PCM encodings where quantization levels vary as a function of amplitude (as with the A-law algorithm or the μ-law algorithm). Though PCM is a more general term, it is often used to describe data encoded as LPCM.

A PCM stream has two basic properties that determine the stream's fidelity to the original analog signal: The sampling rate, which is the number of times per second that samples are taken; and the bit depth, which determines the number of possible digital values that can be used to represent each sample.

1. Implementations

PCM is the method of encoding generally used for uncompressed audio, although there are other methods such as pulse-density modulation (PDM)[1] (used also on Super Audio CD).

The 4ESS switch introduced time-division switching into the U. S. telephone system in 1976, based on medium scale integrated circuit technology.

LPCM is used for the lossless encoding of audio data in the Compact disc Red Book standard (informally also known as Audio CD), introduced in 1982.

AES3 (specified in 1985, upon which S/PDIF is based) is a particular format using LPCM.

On PCs, PCM and LPCM often refer to the format used in WAV (defined in 1991) and AIFF audio container formats (defined in 1988). LPCM data may also be stored in other formats such as AU, raw audio format (header-less file) and various multimedia container formats.

LPCM has been defined as a part of the DVD (since 1995) and Blu-ray (since 2006) standards. It is also defined as a part of various digital video and audio storage formats (e. g. DV since 1995, AVCHD[2] since 2006).

LPCM is used by HDMI (defined in 2002), a single-cable digital audio/video connector interface for transmitting uncompressed digital data.

RF64 container format (defined in 2007) uses LPCM and also allows non-PCM bitstream storage: Various compression formats contained in the RF64 file as data bursts (Dolby E, Dolby AC3, DTS, MPEG-1/MPEG-2 Audio) can be "disguised" as PCM linear.

2. Modulations

In the Figure 10-1, a sine wave (curve) is sampled and quantized for PCM. The sine wave is

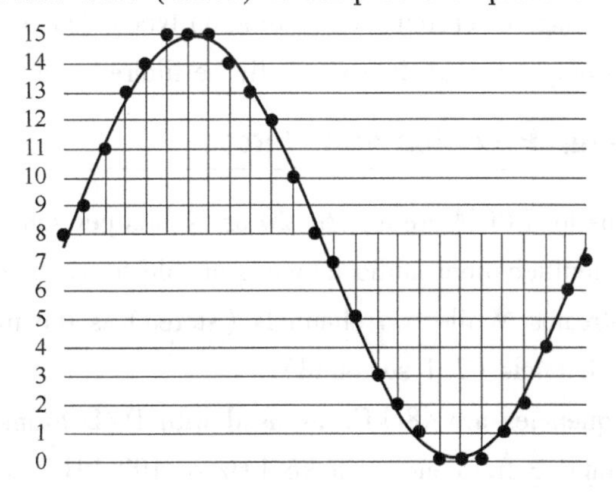

Figure 10-1

sampled at regular intervals, shown as vertical lines. For each sample, one of the available values (on the y-axis) is chosen by some algorithm. This produces a fully discrete representation of the input signal (points) that can be easily encoded as digital data for storage or manipulation. For the sine wave example at right, we can verify that the quantized values at the sampling moments are 8, 9, 11, 13, 14, 15, 15, 15, 14, etc. Encoding these values as binary numbers would result in the following set of nibbles: 1000 ($2^3 \times 1 + 2^2 \times 0 + 2^1 \times 0 + 2^0 \times 0 = 8 + 0 + 0 + 0 = 8$), 1000, 1001, 1011, 1101, 1110, 1111, 1111, 1111, 1110, etc. These digital values could then be further processed or analyzed by a digital signal processor. Several PCM streams could also be multiplexed into a larger aggregate data stream, generally for transmission of multiple streams over a single physical link. One technique is called time-division multiplexing (TDM) and is widely used, notably in the modern public telephone system.

The PCM process is commonly implemented on a single integrated circuit generally referred to as an analog-to-digital converter (ADC).

3. Demodulation

To recover the original signal from the sampled data, a "demodulator" can apply the procedure of modulation in reverse. After each sampling period, the demodulator reads the next value and shifts the output signal to the new value. As a result of these transitions, the signal has a significant amount of high-frequency energy caused by aliasing. To remove these undesirable frequencies and leave the original signal, the demodulator passes the signal through analog filters that suppress energy outside the expected frequency range (greater than the Nyquist[3] frequency). The sampling theorem shows PCM devices can operate without introducing distortions within their designed frequency bands if they provide a sampling frequency twice that of the input signal. For example, in telephony, the usable voice frequency band ranges from approximately 300 Hz to 3400 Hz. Therefore, per the Nyquist-Shannon sampling theorem, the sampling frequency (8 kHz) must be at least twice the voice frequency (4 kHz) for effective reconstruction of the voice signal.

The electronics involved in producing an accurate analog signal from the discrete data are similar to those used for generating the digital signal. These devices are Digital-to-analog converters (DACs). They produce a voltage or current (depending on type) that represents the value presented on their digital inputs. This output would then generally be filtered and amplified for use.

4. Standard Sampling Precision and Rates

Common sample depths for LPCM are 8, 16, 20 or 24 bits per sample. LPCM encodes a single sound channel. Support for multichannel audio depends on file format and relies on interweaving or synchronization of LPCM streams. While two channels (stereo) is the most common format, some can support up to 8 audio channels (7.1 surround).

Common sampling frequencies are 48 kHz as used with DVD format videos, or 44.1 kHz as used in Compact discs. Sampling frequencies of 96 kHz or 192 kHz can be used on some newer equipment, with the higher value equating to 6.144 megabit per second for two channels at 16-bit per sample value, but the benefits have been debated. The bitrate limit for LPCM audio on DVD-Video is also 6.144 Mbit/s, allowing 8 channels (7.1 surround) × 48 kHz × 16-bit per sample = 6, 144 kbit/s. There is a L32 bit PCM, and there are many sound cards that support it.

New Words

compact	[kəm'pækt]	*adj.* 紧凑的，紧密的；简洁的
audio	['ɔːdɪəʊ]	*adj.* 声音的；[声] 音频的，[声] 声频的
application	[ˌæplɪ'keɪʃən]	*n.* 应用；应用程序；应用软件
fidelity	[fɪ'delɪti]	*n.* 保真度；忠诚；精确；尽责
density	['densɪti]	*n.* 密度
lossless	['lɒslɪs]	*adj.* 无损的
container	[kən'teɪnə]	*n.* 集装箱；容器
sine	[saɪn]	*n.* 正弦
verify	['verɪfaɪ]	*v.* 核实；查证
algorithm	['ælgərɪð(ə)m]	*n.* [计] [数] 算法，运算法则
discrete	[dɪ'skriːt]	*n.* 分立元件；独立部件　*adj.* 离散的，不连续的
nibble	['nɪb(ə)l]	*n.* 轻咬；啃；细咬
notably	['nəʊtəbli]	*adv.* 显著地；尤其
interweave	[ˌɪntə'wiːv]	*v.* （使）交织；织进；（使）混杂
format	['fɔːmæt]	*n.* 格式；版式；开本　*v.* 使格式化

Notes

[1] pulse density modulation（PDM，脉冲密度调制）是一种使用二进制数 0、1 表示模拟信号的调制方式。在 PDM 信号中，模拟信号的幅值使用输出脉冲对应区域的密度表示。

[2] AVCHD 是索尼（Sony）公司与松下电器（Panasonic）于 2006 年 5 月联合发布的高画质光碟压缩技术，AVCHD 标准基于 MPEG-4 AVC/H.264 视讯编码，支持 480i、720p、1080i、1080p 等格式，同时支持杜比数位 5.1 声道 AC-3 或线性 PCM 7.1 声道音频压缩。

[3] Nyquist：奈奎斯特（1889—1976），美国物理学家。1917 年获得耶鲁大学工学博士学位。曾在美国 AT&T 公司与贝尔实验室任职。奈奎斯特为近代信息理论做出了突出贡献。他总结的奈奎斯特采样定理是信息论、特别是通信与信号处理学科中的一个重要基本结论。

Questions for Discussion

1. What is the sampling rate?
2. What is the bit depth?
3. How does a demodulator apply to recover the original signal from the sampled data?

Unit 11

Text A

How WiFi Works

If you've been in an airport, coffee shop, library or hotel recently, chances are you've been right in the middle of a wireless network. Many people also use wireless networking, also called WiFi or 802. 11 networking[1], to connect their computers at home, and some cities are trying to use the technology to provide free or low-cost Internet access to residents. In the near future, wireless networking may become so widespread that you can access the Internet just about anywhere at any time, without using wires.

WiFi has a lot of advantages. Wireless networks are easy to set up and inexpensive. They're also unobtrusive—unless you're on the lookout for a place to watch streaming movies on your tablet, you may not even notice when you're in a hotspot. In this article, we'll look at the technology that allows information to travel over the air. We'll also review what it takes to create a wireless network in your home.

1. What Is WiFi?

A wireless network uses radio waves, just like cell phones, televisions and radios do. In fact, communication across a wireless network is a lot like two-way radio communication. Here's what happens:

1) A computer's wireless adapter translates data into a radio signal and transmits it using an antenna.

2) A wireless router receives the signal and decodes it. The router sends the information to the Internet using a physical, wired Ethernet connection.

The process also works in reverse, with the router receiving information from the Internet, translating it into a radio signal and sending it to the computer's wireless adapter.

The radios used for WiFi communication are very similar to the radios used for walkie-talkies, cell phones and other devices. They can transmit and receive radio waves, and they can convert 1s and 0s into radio waves and convert the radio waves back into 1s and 0s. But WiFi radios have a few notable differences from other radios:

1) They transmit at frequencies of 2.4 GHz or 5 GHz. This frequency is considerably higher than the frequencies used for cell phones, walkie-talkies and televisions. The higher frequency allows the signal to carry more data.

2) They use 802.11 networking standards, which come in several flavors:

① 802.11a transmits at 5 GHz and can move up to 54 megabits of data per second. It also uses orthogonal frequency-division multiplexing (OFDM), a more efficient coding technique that splits that radio signal into several sub-signals before they reach a receiver. This greatly reduces interference.

② 802.11b is the slowest and least expensive standard. For a while, its cost made it popular, but now it's becoming less common as faster standards become less expensive. 802.11b transmits in the 2.4 GHz frequency band of the radio spectrum. It can handle up to 11 megabits of data per second, and it uses complementary code keying (CCK) modulation to improve speeds.

③ 802.11g transmits at 2.4 GHz like 802.11b, but it's a lot faster—it can handle up to 54 megabits of data per second. 802.11g is faster because it uses the same OFDM coding as 802.11a.

④ 802.11n is the most widely available of the standards and is backward compatible with a, b and g. It significantly improved speed and range over its predecessors. For instance, although 802.11g theoretically moves 54 megabits of data per second, it only achieves real-world speeds of about 24 megabits of data per second because of network congestion. 802.11n, however, reportedly can achieve speeds as high as 140 megabits per second. 802.11n can transmit up to four streams of data, each at a maximum of 150 megabits per second, but most routers only allow for two or three streams.

⑤ 802.11ac is the newest standard as of early 2013. It has yet to be widely adopted, and is still in draft form at the Institute of Electrical and Electronics Engineers (IEEE), but devices that support it are already on the market. 802.11ac is backward compatible with 802.11n (and therefore the others, too), with n on the 2.4 GHz band and ac on the 5 GHz band. It is less prone to interference and far faster than its predecessors, pushing a maximum of 450 megabits per second on a single stream, although real-world speeds may be lower. Like 802.11n, it allows for transmission on multiple spatial streams—up to eight, optionally. It is sometimes called 5G WiFi because of its frequency band, sometimes Gigabit WiFi because of its potential to exceed a gigabit per second on multiple streams and sometimes Very High Throughput (VHT) for the same reason.

⑥ Other 802.11 standards focus on specific applications of wireless networks, like wide area networks (WANs) inside vehicles or technology that lets you move from one wireless network to another seamlessly.

3) WiFi radios can transmit on any of three frequency bands. Or, they can "frequency hop" rapidly between the different bands. Frequency hopping helps reduce interference and lets multiple devices use the same wireless connection simultaneously.

As long as they all have wireless adapters, several devices can use one router to connect to the

Internet. This connection is convenient, virtually invisible and fairly reliable; however, if the router fails or if too many people try to use high-bandwidth applications at the same time, users can experience interference or lose their connections. Although newer, faster standards like 802.11ac could help with that.

Next, we'll look at how to connect to the Internet from a WiFi hotspot.

2. WiFi Hotspots

A WiFi hotspot is simply an area with an accessible wireless network. The term is most often used to refer to wireless networks in public areas like airports and coffee shops. Some are free and some require fees for use, but in either case they can be handy when you are on the go. You can even create your own mobile hotspot using a cell phone or an external device that can connect to a cellular network. And you can always set up a WiFi network at home.

If you want to take advantage of public WiFi hotspots or your own home-based network, the first thing you'll need to do is make sure your computer has the right gear. Most new laptops and many new desktop computers come with built-in wireless transmitters, and just about all mobile devices are WiFi enabled. If your computer isn't already equipped, you can buy a wireless adapter that plugs into the PC card slot or USB port. Desktop computers can use USB adapters, or you can buy an adapter that plugs into the PCI slot inside the computer's case. Many of these adapters can use more than one 802.11 standard.

Once you've installed a wireless adapter and the drivers that allow it to operate, your computer should be able to automatically discover existing networks. This means that when you turn your computer on in a WiFi hotspot, the computer will inform you that the network exists and ask whether you want to connect to it. If you have an older computer, you may need to use a software program to detect and connect to a wireless network.

Being able to connect to the Internet in public hotspots is extremely convenient. Wireless home networks are convenient as well. They allow you to easily connect multiple computers and to move them from place to place without disconnecting and reconnecting wires.

3. Building a Wireless Network

If you already have several computers networked in your home, you can create a wireless network with a wireless access point. If you have several computers that are not networked, or if you want to replace your Ethernet network, you'll need a wireless router. This is a single unit that contains:

- A port to connect to your cable or DSL modem;
- A router;
- An Ethernet hub[2];
- A firewall;
- A wireless access point.

A wireless router allows you to use wireless signals or Ethernet cables to connect your computers and mobile devices to one another, to a printer and to the Internet. Most routers provide coverage for about 100 feet (30.5 meters) in all directions, although walls and doors can block the signal. If your home is very large, you can buy inexpensive range extenders or repeaters to increase your router's range.

Once you plug in your router, it should start working at its default settings. Most routers let you use a Web interface to change your settings. You can select:

- The name of the network, known as its service set identifier (SSID)—The default setting is usually the manufacturer's name.
- The channel that the router uses—Most routers use channel 6 by default. If you live in an apartment and your neighbors are also using channel 6, you may experience interference. Switching to a different channel should eliminate the problem.
- Your router's security options—Many routers use a standard, publicly available sign-on, so it's a good idea to set your own username and password.

Security is an important part of a home wireless network, as well as public WiFi hotspots. If you set your router to create an open hotspot, anyone who has a wireless card will be able to use your signal. Most people would rather keep strangers out of their network, though. Doing so requires you to take a few security precautions.

It's also important to make sure your security precautions are current. The wired equivalency privacy (WEP) security measure was once the standard for WAN security. The idea behind WEP was to create a wireless security platform that would make any wireless network as secure as a traditional wired network. But hackers discovered vulnerabilities in the WEP approach, and today it's easy to find applications and programs that can compromise a WAN running WEP security. It was succeeded by the first version of WiFi Protected Access (WPA), which uses Temporal Key Integrity Protocol (TKIP)[3] encryption and is a step up from WEP, but is also no longer considered secure.

To keep your network private, you can use one or both of the following methods:

- WiFi Protected Access version 2 (WPA2) is the successor to WEP and WPA, and is now the recommended security standard for WiFi networks.
- Media Access Control (MAC) address filtering is a little different from WEP, WPA or WPA2. It doesn't use a password to authenticate users—it uses a computer's physical hardware.

Wireless networks are easy and inexpensive to set up, and most routers' Web interfaces are virtually self-explanatory.

New Words

unobtrusive	[ˈʌnəbˈtruːsɪv]	*adj.* 不突出的，不引人注目的；不唐突的
notable	[ˈnəʊtəb(ə)l]	*adj.* 值得注意的；显著的；著名的
flavor	[ˈfleɪvə]	*n.* 特点；韵味；味
handle	[ˈhænd(ə)l]	*v.* 处理；操作；运用；买卖；触摸

improve	[ɪm'pruːv]	*v.* 改善，增进；提高……的价值
significantly	[sɪg'nɪfɪk(ə)ntli]	*adv.* 显著地；相当数量地
predecessor	['priːdɪsesə]	*n.* 前身，原有事物；前任，前辈
router	['ruːtə (r)]	*n.* 路由器（传送信息的专用网络智能设备）
congestion	[kən'dʒestʃ(ə)n]	*n.* 阻塞；拥挤，堵车
extender	[ɪk'stendə]	*n.* 扩充器；延长器
repeater	[rɪ'piːtə]	*n.* [通信] 中继器；[通信] 转发器
adapter	[ə'dæptə]	*n.* 适配器；改编者；接合器
default	[dɪ'fɔːlt; 'diːfɔːlt]	*n.* 系统默认值；违约；缺乏
hotspot	['hɒtspɒt]	*n.* 热点；热区
equivalency	[ɪ'kwɪvələnsi]	*n.* 等价；相等（等于 equivalence）

Notes

[1] 802.11 协议簇是国际电工电子工程学会（IEEE）为无线局域网络制定的标准。虽然 WiFi 使用了 802.11 的媒体访问数据链路层（DLL）和物理层（PHY），但是两者并不完全一致。在一些标准中，使用最多的应该是 802.11n 标准，工作在 2.4GHz 或 5GHz 频段，可达 600Mbit/s（理论值）。IEEE 最初制定的一个无线局域网标准，主要用于解决办公室局域网和校园网中用户与用户终端的无线接入，业务主要限于数据存取，速率最高只能达到 2Mbit/s。由于它在速率和传输距离上都不能满足人们的需要，因此 IEEE 小组又相继推出了 802.11b 和 802.11a 两个新标准。

[2] hub 是一个多端口的转发器，在以 HUB 为中心设备时，即使网络中某条线路产生了故障，也不影响其他线路的工作，所以 HUB 在局域网中得到了广泛的应用。大多数时候它用在星型与树型网络拓扑结构中，以 RJ45 接口与各主机相连（也有 BNC 接口），HUB 按照不同的说法有很多种类。HUB 按照对输入信号的处理方式，可以分为无源 HUB、有源 HUB 和智能 HUB。

[3] Temporal Key Integrity Protocol（TKIP）是 IEEE802.11i 规范中负责处理无线安全问题的加密协议。在 IEEE 802.11i 规范中，TKIP 负责处理无线安全问题的加密部分。TKIP 实现了密钥混合功能，结合秘密与根密钥初始化向量传递给它的 RC4 初始化之前。TKIP 是包裹在已有 WEP 密码外围的一层“外壳”。TKIP 由 WEP 使用的同样的加密引擎和 RC4 算法组成。

Questions for Discussion

1. What is WiFi?
2. What does WiFi Hotspots mean?
3. How to set up a WiFi connection in your home?

Text B

NFC: Getting Down to Business

—by Koichi Tagawa, Chairman, NFC Forum

The year 2013 marks an inflection point in the global adoption of Near Field Communication (NFC)[1] technology. As of this year, many of the key components considered essential to the advancement of NFC—a comprehensive set of specifications, NFC-enabled device availability, a robust certification program—are now in place. For those of us who have dedicated the last decade to bringing NFC to the world, it is exciting to realize that "someday" has finally become "today". In recognition of this inflection point, the NFC Forum has begun to expand its efforts to better support the implementation of NFC solutions in key vertical industries and market segments. The signs of accelerating progress can be seen in several areas:

1. NFC-Enabled Devices of All Kinds Are Now Widely Available

According to ABI Research, over 100 million NFC devices shipped in 2012, and there will be close to 300 million shipped this year. It's important to note that the list of commercially-available NFC-enabled devices is not only long, but also diverse—spanning everything from smart phones and tablets to gaming consoles, laptops, speakers, and even washing machines. Most consumers seeking NFC-enabled devices can find them; many consumers who know little or nothing about NFC will get NFC in the next smartphone or tablet they purchase, whether they realize it or not.

2. NFC Forum Specifications Are Supporting More Functionality and Market Needs

Last October, the NFC Forum approved and adopted the NFC Analog technical specification. This marked a major step toward our goal of global interoperability by making it easier for device manufacturers to build NFC Forum-compliant devices. Because it addresses the analog characteristics of the RF interface of an NFC-enabled device, it also streamlines testing and certification, which is why we expect that it will accelerate the introduction of NFC-enabled devices into the market.

A month later, we published the NFC Controller Interface (NCI) technical specification, a major new specification that defines a standard interface within an NFC device between an NFC controller and the device's main application processor. The availability of the NCI specification is significant because it makes it easier for device manufacturers to integrate chipsets from different chip manufacturers, and it defines a common level of functionality and interoperability among the components within an NFC-enabled device. With the availability of the NCI, manufacturers have access to a standard interface they can use for whatever kind of NFC-enabled device they build—including mobile phones, PCs, tablets, printers, consumer electronics, and appliances. This will ease chip sourcing and, again, reduce time to market for new NFC-enabled devices of all kinds. More recently, we announced the new Personal Health Device Communication (PHDC) specification—our first specification supporting a specific vertical market—health care. We also announced important updates to the Connection Handover (release 1.3) and Signature RTD (release 2.0) specifications.

3. The NFC Forum Certification Program Has Been Expanded

For device manufacturers seeking to ensure that their products conform to NFC Forum specifications, certification is essential. We've enhanced our NFC Forum Certification Program to be more robust. Device manufacturers can now test their products against the latest versions of the Digital Protocol, Tag Type Operations, LLCP, SNEP[2], and Analog specifications, thus providing added confidence and assurance for companies bringing new NFC devices to market.

4. NFC Solutions Are Popping Up Everywhere

One of the most gratifying things about working on NFC technology is to see the remarkable and growing diversity of NFC solutions being brought to market. For example:

1) One of the five largest automakers in the world recently announced that it will offer a solution that allows NFC enabled devices to integrate with its cars—for keyless access, automatic personalized settings, music play, and more—in 2015.

2) Late last year, a top-three video game manufacturer introduced a new console that has NFC built into its controller.

3) A leading game publisher introduced an NFC-enabled video game that quickly became the #1 children's game of 2012.

4) A top location-based social network now supports NFC on major smart phone platforms, providing 35 million users and 1.4 million businesses with the ability to more easily share recommendations and deals all over the world.

5) The largest statewide public transportation system in America is averaging 10,000 NFC-enabled mobile payment transactions per month for rail and bus fares.

6) A Japanese online retailer allows shoppers to use NFC to tap product displays and make online purchases for home delivery.

7) A Swedish/American joint venture has launched a small NFC-enabled ECG device that tracks heart arrhythmia data of atrial fibrillation patients for transmission via NFC to their physicians for monitoring and management.

8) And last but not least, my company, Sony, has built NFC into a growing number of wireless speakers, PCs, smart phones, and other media devices for fast and easy Bluetooth and WiFi pairing.

What Do These Developments Mean?

These are just a few examples of companies that have made major strategic commitments to NFC. The point is that successful innovation is contagious. When an NFC-enabled video game can become the global top seller, winning "Game of the Year" and "Innovative Toy of the Year" awards from the Toy Industry Association, you can be sure we'll continue to see other games that take advantage of NFC technology. When a carmaker makes the strategic decision to use NFC as the means to integrate and personalize customers' driving experiences with their mobile devices, other manufacturers will likely follow suit (and in fact, already are). And when NFC support is considered essential to a social network's growth plans, there is little question that newer startups will make the same strategic decision.

In all of these examples, the companies determined that the added value of NFC was clear, significant, and good for business. They need no further convincing that NFC can help them gain a competitive advantage and grow their revenue. Other companies will come to the same conclusion—if they're fully informed.

What's Needed Now?

Early-adopter companies are essential to proving the viability of commercial NFC solutions. Ultimately, however, widespread adoption and use of NFC will depend on consumer demand. For that demand to grow, people need to know what NFC is and what it can do for them. Since NFC-enabled mobile payments have garnered significant media coverage, many consumers who are already aware of NFC may think of it only as a mobile payment technology. It's true that mobile payment is a huge opportunity for NFC. According to the latest forecast from Transparency Market Research, the global mobile wallet market is expected to reach $ 1, 602. 4 billion by 2018, with much of this market success attributed to NFC. However, we in the NFC ecosystem are not serving the needs of the marketplace if all we do is wait for the NFC payment business to mature, providing the coattails for other NFC services to riding on. It's important for consumers with NFC-enabled devices today to know that NFC technology has many, many other use cases that can make an immediate, positive impact on their lives.

Getting that message across will require a series of actions: We need to increase consumer awareness of what NFC is and what it can do for people. Although hundreds of millions of NFC-enabled devices are entering the marketplace, few of them are packaged with instructions on the capabilities and uses of NFC, or a way for consumers to quickly experience NFC in action. While it is gratifying to see a major smart phone provider making NFC file-sharing the centerpiece of several television commercials, manufacturers should consider other ways to generate awareness and encourage consumer use of NFC, including providing:

1) Samples of NFC tags in smart phone packaging for consumers to try out;

2) Preloaded applications that use NFC.

We need more NFC services. The number and variety of NFC solutions continue to grow, but commercial NFC services for consumers still remain relatively scarce. Just as tablet users who become used to touch screen interfaces often get frustrated when they go back to PC menus and mouse clicks, consumers who become accustomed to NFC's intuitive touch-and-go interface want to be able to perform more and more actions with a quick tap of their devices. Solutions providers across industries need to be ready to satisfy that need. To give the introduction of new NFC services a boost, the NFC Forum continues to clear the path to commercialization with new specifications that lower barriers to market entry, solutions showcases that promote new ideas, and plug fests that bring developers together to test their products for interoperability.

Every solutions developer should have an NFC strategy. A decade ago, many solutions developers were slow to see how mobile devices would affect their products and their use. Those who did benefited from being first to market with mobile versions or apps that gave users easier access, greater convenience, more information, or new capabilities. NFC offers similar opportunities. Evans Data Corporation's recently released Mobile Development Survey states that over 31% of mobile developers today are supporting NFC in their mobile applications. While this is an encouraging sign, the other 69% should already be evaluating how they can better meet customer needs and advance their business strategy by taking advantage of NFC.

Businesses need to explore how NFC can drive new growth. The first mobile "solutions" were little more than conventional websites adapted to display properly on smaller mobile phone screens. However, within a few years, mobile became a market segment unto itself, launching a new wave of innovation that is still on the rise. NFC has that same transformative power—in fact, even more so, because NFC goes beyond mobile, enabling links between the Internet and everyday devices and objects, from household appliances to store signage. As greater intelligence is embedded in these everyday objects, NFC has the power to put that intelligence to work in new ways—for both internal business and external customer applications.

For example, a company in France developed an NFC-enabled solution that monitors wine shipment temperatures across distribution channels to ensure the wine's provenance and quality. Each box of wine is equipped with a battery-powered RFID[3] temperature sensor. At each step of the distribution cycle, the wine can be checked for temperature and authenticity using an NFC-enabled device. Without NFC, this solution would have been costly and impractical. The business advantages of NFC are many. Companies with mobile workforces, such as traveling service personnel, can better track and direct their actions with NFC. Marketers, advertisers, and retailers seeking to build 1 : 1 marketing relationships with valued consumers can leverage NFC for personalized offers where they have the greatest impact at point of sale.

New Words

inflection [ɪnˈflekʃ(ə)n] *n.* 弯曲，变形；音调变化

essential [ɪˈsenʃ(ə)l] *n.* 本质；要素 *adj.* 基本的；必要的

availability [əˌveɪləˈbɪləti] *n.* 可用性；有效性；实用性

vertical [ˈvɜːtɪk(ə)l] *n.* 垂直线，垂直面 *adj.* 垂直的，直立的

purchase [ˈpɜːtʃəs] *n.* 购买；紧握 *v.* 购买，赢得；购买东西

interoperability [ˈɪntərˌɒpərəˈbɪləti] *n.* [计] 互操作性；互用性

interface [ˈɪntəfeɪs] *n.* <计>接口；交界面 *v.* [计算机] 使联系

gratify [ˈgrætɪfaɪ] *v.* 使满足；使满意，使高兴

innovation [ˌɪnəˈveɪʃn] *n.* 创新，革新；新方法

contagious	[kən'teɪdʒəs]	*adj.* 感染性的；会蔓延的
revenue	['revənjuː]	*n.* 税收，国家的收入；收益
viability	[ˌvaɪə'bɪləti]	*n.* 生存能力，发育能力；可行性
frustrate	[frʌ'streɪt]	*v.* 挫败；阻挠；使感到灰心
signage	['saɪnɪdʒ]	*n.* 引导标示

Notes

[1] NFC 近场通信技术是由非接触式射频识别（RFID）及互联互通技术整合演变而来的，在单一芯片上结合感应式读卡器、感应式卡片和点对点的功能，能在短距离内与兼容设备进行识别和数据交换。工作频率为 13.56MHz，但是使用这种手机支付方案的用户必须更换特制的手机。目前这项技术在日韩被广泛应用，他们的手机可以用作机场登机验证、大厦的门禁钥匙、交通一卡通、信用卡、支付卡等。

[2] SPINS 安全协议框架是最早的无线传感网络（Wireless Sensor NetWork，WSN）安全框架之一，包括 SNEP（Secure Network Encryption Protocol）和 μTESLA（micro Timed Efficient Streaming Loss 2 tolerant Authentication Protocol）两个部分。

[3] RFID（Radio Frequency Identification）技术，又称无线射频识别，是一种通信技术，可通过无线电信号识别特定目标并读写相关数据，而无须在识别系统与特定目标之间建立机械或光学接触。

Questions for Discussion

1. How can device manufacturers be more confident when bringing new NFC devices to market?
2. How does Sony apply NFC?
3. What is the NFC-enabled solution applied in the French company?

Unit 12

Text A

Digital Signal Processing

Digital signal processing (DSP) is the use of digital processing, such as by computers, to perform a wide variety of signal processing operations. The signals processed in this manner are a sequence of numbers that represent samples of a continuous variable in a domain such as time, space, or frequency.

Digital signal processing and analog signal processing are subfields of signal processing. DSP applications include audio and speech signal processing, sonar, radar and other sensor array processing, spectral estimation, statistical signal processing, digital image processing, signal processing for telecommunications, control of systems, biomedical engineering, seismic data processing, among others.

Digital signal processing can involve linear or nonlinear operations. Nonlinear signal processing is closely related to nonlinear system identification and can be implemented in the time, frequency, and spatio-temporal domains.

The application of digital computation to signal processing allows for many advantages over analog processing in many applications, such as error detection and correction in transmission as well as data compression. DSP is applicable to both streaming data and static (stored) data (see Figure 12-1).

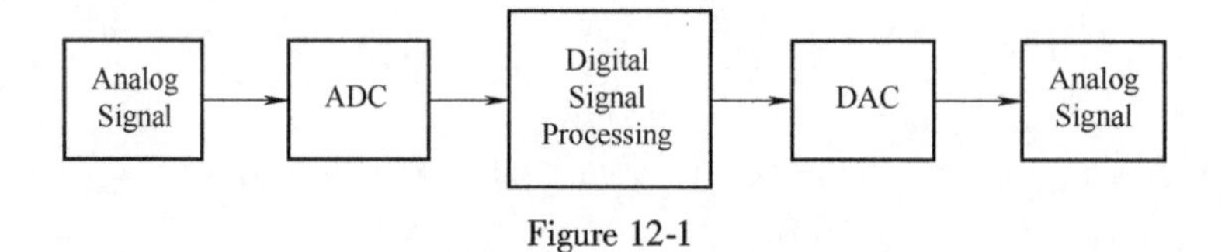

Figure 12-1

1. Signal Sampling

The increasing use of computers has resulted in the increased use of, and need for, digital signal processing. To digitally analyze and manipulate an analog signal, it must be digitized with an analog-to-digital converter. Sampling is usually carried out in two stages: discretization and quantization. Discretization means that the signal is divided into equal intervals of time, and each interval is represented by a single measurement of amplitude. Quantization means each amplitude measurement is approximated by a value from a finite set. Rounding real numbers to integers is an example.

The Nyquist Shannon sampling theorem states that a signal can be exactly reconstructed from its samples if the sampling frequency is greater than twice the highest frequency of the signal, but this requires an infinite number of quantization levels. In practice, the sampling frequency is often significantly higher than twice that required by the signal's limited bandwidth.

Theoretical DSP analyses and derivations are typically performed on discrete-time signal models with no amplitude inaccuracies (quantization error), "created" by the abstract process of sampling. Numerical methods require a quantized signal, such as those produced by an analog-to-digital converter (ADC)[1]. The processed result might be a frequency spectrum or a set of statistics. But often it is another quantized signal that is converted back to analog form by a digital-to-analog converter (DAC).

2. Domains

In DSP, engineers usually study digital signals in one of the following domains: time domain (one-dimensional signals), spatial domain (multidimensional signals), frequency domain, and wavelet domains. They choose the domain in which to process a signal by making an informed assumption (or by trying different possibilities) as to which domain best represents the essential characteristics of the signal. A sequence of samples from a measuring device produces a temporal or spatial domain representation, whereas a discrete Fourier transform produces the frequency domain information, that is, the frequency spectrum.

3. Time and Space Domains

The most common processing approach in the time or space domain is enhancement of the input signal through a method called filtering. Digital filtering generally consists of some linear transformation of a number of surrounding samples around the current sample of the input or output signal. There are various ways to characterize filters; for example:

1) A "linear" filter is a linear transformation of input samples; other filters are "non-linear". Linear filters satisfy the superposition condition, i. e. if an input is a weighted linear combination of different signals, the output is a similarly weighted linear combination of the corresponding output signals.

2) A "causal" filter uses only previous samples of the input or output signals; while a "non-causal" filter uses future input samples. A non-causal filter can usually be changed into a causal filter by adding a delay to it.

3) A "time-invariant" filter has constant properties over time; other filters such as adaptive

filters change in time.

4) A "stable" filter produces an output that converges to a constant value with time, or remains bounded within a finite interval. An "unstable" filter can produce an output that grows without bounds, with bounded or even zero input.

5) A "finite impulse response" (FIR) filter uses only the input signals, while an "infinite impulse response" filter (IIR) uses both the input signal and previous samples of the output signal. FIR filters are always stable, while IIR filters may be unstable.

A filter can be represented by a block diagram, which can then be used to derive a sample processing algorithm to implement the filter with hardware instructions. A filter may also be described as a difference equation, a collection of zeroes and poles or, if it is an FIR filter, an impulse response or step response.

The output of a linear digital filter to any given input may be calculated by convolving the input signal with the impulse response.

4. Frequency Domain

Signals are converted from time or space domain to the frequency domain usually through the Fourier transform. The Fourier transform converts the signal information to a magnitude and phase component of each frequency. Often the Fourier transform is converted to the power spectrum, which is the magnitude of each frequency component squared.

The most common purpose for analysis of signals in the frequency domain is analysis of signal properties. The engineer can study the spectrum to determine which frequencies are present in the input signal and which are missing.

In addition to frequency information, phase information is often needed. This can be obtained from the Fourier transform. With some applications, how the phase varies with frequency can be a significant consideration.

Filtering, particularly in non-realtime work can also be achieved by converting to the frequency domain, applying the filter and then converting back to the time domain. This is a fast, $O(n \log n)$ operation, and can give essentially any filter shape including excellent approximations to brickwall filters.

There are some commonly used frequency domain transformations. For example, the cepstrum converts a signal to the frequency domain through Fourier transform, takes the logarithm, then applies another Fourier transform. This emphasizes the harmonic structure of the original spectrum. Frequency domain analysis is also called spectrum or spectral analysis.

5. Z-plane Analysis

Digital filters come in both IIR and FIR types. FIR filters have many advantages, but are computationally more demanding. Whereas FIR filters are always stable, IIR filters have feedback loops that may resonate when stimulated with certain input signals. The Z-transform provides a tool for analyzing potential stability issues of digital IIR filters. It is analogous to the Laplace transform, which is used to design analog IIR filters.

6. Wavelet

In numerical analysis and functional analysis, a discrete wavelet transform (DWT) is any wavelet transform for which the wavelets are discretely sampled. As with other wavelet transforms, a key advantage it has over Fourier transforms is temporal resolution: it captures both frequency and location information.

7. Applications

The main applications of DSP are audio signal processing, audio compression, digital image processing, video compression, speech processing, speech recognition, digital communications, digital synthesizers, radar, sonar, financial signal processing, seismology and biomedicine. Specific examples are speech compression and transmission in digital mobile phones, room correction of sound in hi-fi and sound reinforcement applications, weather forecasting, economic forecasting, seismic data processing, analysis and control of industrial processes, medical imaging such as CAT scans and MRI, MP3 compression, computer graphics, image manipulation, hi-fi loudspeaker crossovers and equalization, and audio effects for use with electric guitar amplifiers.

8. Implementation

DSP algorithms have long been run on general-purpose computers and digital signal processors. DSP algorithms are also implemented on purpose-built hardware such as application-specific integrated circuit (ASICs)[2]. Additional technologies for digital signal processing include more powerful general purpose microprocessors, field-programmable gate arrays (FPGAs)[3], digital signal controllers (mostly for industrial applications such as motor control), and stream processors.

Depending on the requirements of the application, digital signal processing tasks can be implemented on general purpose computers.

Often when the processing requirement is not real-time, processing is economically done with an existing general-purpose computer and the signal data (either input or output) exists in data files. This is essentially no different from any other data processing, except DSP mathematical techniques (such as the FFT) are used, and the sampled data is usually assumed to be uniformly sampled in time or space. For example: processing digital photographs with software such as Photoshop.

However, when the application requirement is real-time, DSP is often implemented using specialized microprocessors such as the DSP56000, the TMS320, or the SHARC. These often process data using fixed-point arithmetic, though some more powerful versions use floating point. For faster applications FPGAs might be used. Beginning in 2007, multicore implementations of DSPs have started to emerge from companies including Freescale[4] and Stream Processors, Inc. For faster applications with vast usage, ASICs might be designed specifically. For slow applications, a traditional slower processor such as a microcontroller may be adequate. Also a growing number of

DSP applications are now being implemented on embedded systems using powerful PCs with multi-core processors.

New Words

sequence	[ˈsiːkw(ə)ns]	*n.* 数列，序列；顺序；连续
domain	[də(ʊ)ˈmeɪn]	*n.* [计] 域名；范围，疆土；管辖范围
subfield	[ˈsʌbfiːld]	*n.* 子域；分栏；子字段；分支
estimation	[estɪˈmeɪʃ(ə)n]	*n.* 估计；评价；判断
seismic	[ˈsaɪzmɪk]	*adj.* 地震的；由地震引起的；震撼世界的
linear	[ˈlɪnɪə]	*adj.* 直线的，线形的
nonlinear	[nɒnˈlɪnɪə]	*adj.* 非线性的
compression	[kəmˈpreʃ(ə)n]	*n.* 压缩，压紧，浓缩，紧缩
discretization	[dɪsˈkriːtaɪˈzeɪʃən]	*n.* 离散化
quantization	[ˌkwɒntaɪˈzeɪʃən]	*n.* 量子化；数字化；量化
approximate	[əˈprɒksɪmət]	*adj.* 近似的 *v.* 接近于；使接近；使结合
rounding	[ˈraʊndɪŋ]	*adj.* 圆的，环绕的，凑整的
integer	[ˈɪntɪdʒə]	*n.* 整数
theorem	[ˈθɪərəm]	*n.* [数] 定理；（能证明的）一般原理，定律
derivation	[derɪˈveɪʃ(ə)n]	*n.* 引出，导出；衍生
quantize	[ˈkwɒntaɪz]	*v.* 使量子化
enhancement	[ɪnˈhɑːnsm(ə)nt]	*n.* 增强；增加；提高；改善
invariant	[ɪnˈveərɪənt]	*adj.* 无变化的，不变的 *n.* 不变式，不变量
adaptive	[əˈdæptɪv]	*adj.* 适应的；有适应能力的
convolve	[kənˈvɒlv]	*v.* 卷，盘旋，缠绕
cepstrum	[sepstˈrʌm]	*n.* 对数倒频谱，对数逆谱，倒频谱
logarithm	[ˈlɒgərɪð(ə)m]	*n.* 对数
capture	[ˈkæptʃə]	*v.* 俘获；夺取；夺得；引起（注意、想象、兴趣）
synthesizer	[ˈsɪnθɪsaɪzə]	*n.* 合成者，合成物；合成器，综合器
arithmetic	[əˈrɪθmətɪk]	*n.* 算术，计算；算法

Notes

[1] ADC：将模拟信号转换成数字信号的电路，称为模数转换器（简称 A/D 转换器或 ADC，analog-to-digital converter）。A/D 转换的作用是将时间连续、幅值也连续的模拟量转换为时间离散、幅值也离散的数字信号，因此，A/D 转换一般要经过取样、保持、量化及编码 4 个过程。在实际电路中，这些过程有的是合并进行的。模数转换器的种类很多，按工作原理的不同，可分成间接 ADC 和直接 ADC。间接 ADC 是先将输入模拟电压转换成时间或频率，然后再把这些中间量转换成数字量，常用的有中间量是时间的双积分型 ADC。直接 ADC 则直接转换成数字量，常用的有并联比较型 ADC 和逐次逼近型 ADC。

[2] application-specific integrated circuit (ASIC)：在集成电路界 ASIC 被认为是一种为专门目的而设计的集成电路，是指应特定用户要求和特定电子系统的需要而设计、制造的集成电路。ASIC 的特点是面向特定用户的需求。ASIC 在批量生产时与通用集成电路相比具有体积更小、功耗更低、可靠性提高、性能提高、保密性增强、成本降低等优点。

[3] FPGA (field-programmable gate array)，即现场可编程门阵列，它是在 PAL、GAL、CPLD 等可编程器件的基础上进一步发展的产物。它是作为专用集成电路（ASIC）领域中的一种半定制电路而出现的，既解决了定制电路的不足，又克服了原有可编程器件门电路数有限的缺点。

[4] Freescale：飞思卡尔半导体（Freescale Semiconductor）是全球领先的半导体公司，全球总部位于美国得克萨斯州的奥斯汀市。专注于嵌入式处理解决方案。飞思卡尔面向汽车、网络、工业和消费电子市场，提供的技术包括微处理器、微控制器、传感器、模拟集成电路和连接。飞思卡尔的一些主要应用和终端市场包括汽车安全、混合动力和全电动汽车、下一代无线基础设施、智能能源管理、便携式医疗器件、消费电器以及智能移动器件等。在全世界拥有多家设计、研发、制造和销售机构。

Questions for Discussion

1. In what aspects can DSP be used?
2. What are discretization and quantization?
3. According to this passage, what does filtering mean? What kinds of filters do you know?
4. What commonly used frequency domain transformations?

Text B

Digital Signal Processors

A digital signal processor (DSP) is a specialized microprocessor (or an SIP block), with its architecture optimized for the operational needs of digital signal processing.

The goal of DSPs is usually to measure, filter and/or compress continuous real-world analog signals. Most general-purpose microprocessors can also execute digital signal processing algorithms successfully, but dedicated DSPs usually have better power efficiency thus they are more suitable in portable devices such as mobile phones because of power consumption constraints. DSPs often use special memory architectures that are able to fetch multiple data and/or instructions at the same time.

1. Overview

Digital signal processing algorithms typically require a large number of mathematical operations to be performed quickly and repeatedly on a series of data samples. Signals (perhaps from audio or video sensors) are constantly converted from analog to digital, manipulated digitally, and then converted back to analog form. Many DSP applications have constraints on latency; that is, for the system to work, the DSP operation must be completed within some fixed time, and deferred (or batch) processing is not viable.

Most general-purpose microprocessors and operating systems can execute DSP algorithms successfully, but are not suitable for use in portable devices such as mobile phones and PDAs because of power efficiency constraints. A specialized digital signal processor, however, will tend to provide a lower-cost solution, with better performance, lower latency, and no requirements for specialized cooling or large batteries.

The architecture of a DSP is optimized specifically for digital signal processing. Most also support some of the features as an applications processor or microcontroller, since signal processing is rarely the only task of a system. Some useful features for optimizing DSP algorithms are outlined below.

2. Architecture

(1) Software

By the standards of general-purpose processors, DSP instruction sets are often highly irregular; while traditional instruction sets are made up of more general instructions that allow them to perform a wider variety of operations, instruction sets optimized for digital signal processing contain instructions for common mathematical operations that occur frequently in DSP calculations. Both traditional and DSP-optimized instruction sets are able to compute any arbitrary operation but an operation that might require multiple ARM or x86 instructions to compute might require only one instruction in a DSP-optimized instruction set.

One implication for software architecture is that hand-optimized assembly-code routines are commonly packaged into libraries for reuse, instead of relying on advanced compiler technologies to handle essential algorithms. Even with modern compiler optimizations hand-optimized assembly code is more efficient and many common algorithms involved in DSP calculations are hand-written in order to take full advantage of the architectural optimizations.

1) Instruction Sets.

- Multiply-accumulates (MACs, including fused multiply-add, FMA) operations;
- Used extensively in all kinds of matrix operations;
- Convolution for filtering;
- Dot product;
- Polynomial evaluation;
- Fundamental DSP algorithms depend heavily on multiply-accumulate performance;
- FIR filters;
- Fast Fourier transform (FFT);
- Instructions to increase parallelism;
- SIMD;
- VLIW;
- Superscalar architecture;

- Specialized instructions for modulo addressing in ring buffers and bit-reversed addressing mode for FFT cross-referencing;
- DSPs sometimes use time-stationary encoding to simplify hardware and increase coding efficiency;
- Multiple arithmetic units may require memory architectures to support several accesses per instruction cycle;
- Special loop controls, such as architectural support for executing a few instruction words in a very tight loop without overhead for instruction fetches or exit testing.

2) Data Instructions

- Saturation arithmetic, in which operations that produce overflows will accumulate at the maximum (or minimum) values that the register can hold rather than wrapping around (maximum + 1 doesn't overflow to minimum as in many general-purpose CPUs, instead it stays at maximum). Sometimes various sticky bits operation modes are available;
- Fixed-point arithmetic is often used to speed up arithmetic processing;
- Single-cycle operations to increase the benefits of pipelining.

3) Program Flow.

- Floating-point unit integrated directly into the datapath;
- Pipelined architecture;
- Highly parallel multiplier-accumulators (MAC units);
- Hardware-controlled looping, to reduce or eliminate the overhead required for looping operations.

(2) Hardware

1) Memory Architecture. DSPs are usually optimized for streaming data and use special memory architectures that are able to fetch multiple data and/or instructions at the same time, such as the Harvard architecture or Modified Von Neumann architecture, which use separate program and data memories (sometimes even concurrent access on multiple data buses).

2) Addressing and Virtual Memory. DSPs frequently use multi-tasking operating systems, but have no support for virtual memory or memory protection. Operating systems that use virtual memory require more time for context switching among processes, which increases latency.

- Hardware modulo addressing;
- Allows circular buffers to be implemented without having to test for wrapping;
- Bit-reversed addressing, a special addressing mode;
- useful for calculating FFTs;
- Exclusion of a memory management unit;
- Memory-address calculation unit.

3. History

Prior to the advent of stand-alone DSP chips discussed below, most DSP applications were implemented using bit-slice processors. The AMD 2901 bit-slice chip with its family of components was a very popular choice. There were reference designs from AMD, but very often the specifics of a particular design were application specific. These bit slice architectures would sometimes include a peripheral multiplier chip. Examples of these multipliers were a series from TRW including the TDC1008 and TDC1010, some of which included an accumulator, providing the requisite multiply-accumulate (MAC) function.

In 1976, Richard Wiggins proposed the Speak & Spell concept to Paul Breedlove, Larry Brantingham, and Gene Frantz at Texas Instrument's Dallas research facility. Two years later in 1978 they produced the first Speak & Spell, with the technological centerpiece being the TMS5100, the industry's first digital signal processor. It was also the first chip to use Linear predictive coding to perform speech synthesis.

In 1978, Intel released the 2920 as an "analog signal processor". It had an on-chip ADC/DAC with an internal signal processor, but it didn't have a hardware multiplier and was not successful in the market. In 1979, AMI released the S2811. It was designed as a microprocessor peripheral, and it had to be initialized by the host. The S2811 was likewise not successful in the market.

In 1980 the first stand-alone, complete DSPs—the NEC μPD7720 and AT&T DSP1—were presented at the International Solid-State Circuits Conference '80. Both processors were inspired by the research in PSTN[1] telecommunications.

The Altamira DX-1 was another early DSP, utilizing quad integer pipelines with delayed branches and branch prediction.

Another DSP produced by Texas Instruments (TI)[2], the TMS32010 presented in 1983, proved to be an even bigger success. It was based on the Harvard architecture, and so had separate instruction and data memory. It already had a special instruction set, with instructions like load-and-accumulate or multiply-and-accumulate. It could work on 16-bit numbers and needed 390 ns for a multiply-add operation. TI is now the market leader in general-purpose DSPs.

About five years later, the second generation of DSPs began to spread. They had 3 memories for storing two operands simultaneously and included hardware to accelerate tight loops, they also had an addressing unit capable of loop-addressing. Some of them operated on 24-bit variables and a typical model only required about 21 ns for an MAC. Members of this generation were for example the AT&T DSP16A or the Motorola 56000.

The main improvement in the third generation was the appearance of application-specific units and instructions in the data path, or sometimes as coprocessors. These units allowed direct hardware acceleration of very specific but complex mathematical problems, like the Fourier-transform or matrix operations. Some chips, like the Motorola MC68356, even included more than one processor core to work in parallel. Other DSPs from 1995 are the TI TMS320C541 or the TMS 320C80.

The fourth generation is best characterized by the changes in the instruction set and the instruction encoding/decoding. SIMD[3] extensions were added, VLIW and the superscalar architecture appeared. As always, the clock-speeds have increased, a 3 ns MAC now became possible.

4. Modern DSPs

Modern signal processors yield greater performance; this is due in part to both technological and architectural advancements like lower design rules, fast-access two-level cache, (E) DMA circuitry and a wider bus system. Not all DSPs provide the same speed and many kinds of signal processors exist, each one of them being better suited for a specific task, ranging in price from about US $ 1.50 to US $ 300. Texas Instruments produces the C6000 series DSPs, which have clock speeds of 1.2 GHz and implement separate instruction and data caches. They also have an 8 MiB 2nd level cache and 64 EDMA channels. The top models are capable of as many as 8000 MIPS (instructions per second), use VLIW[4] (very long instruction word), perform eight operations per clock-cycle and are compatible with a broad range of external peripherals and various buses (PCI/serial/etc). TMS320C6474 chips each have three such DSPs, and the newest generation C6000 chips support floating point as well as fixed point processing.

Freescale produces a multi-core DSP family—the MSC81xx. The MSC81xx is based on StarCore Architecture processors and the latest MSC8144 DSP combines four programmable SC3400 StarCore DSP cores. Each SC3400 StarCore DSP core has a clock speed of 1 GHz.

XMOS produces a multi-core multi-threaded line of processor well suited to DSP operations. They come in various speeds ranging from 400 to 1600 MIPS. The processors have a multi-threaded architecture that allows up to 8 real-time threads per core, meaning that a 4 core device would support up to 32 real time threads. Threads communicate between each other with buffered channels that are capable of up to 80 Mbit/s. The devices are easily programmable in C and aim at bridging the gap between conventional micro-controllers and FPGAs.

CEVA, Inc. produces and licenses three distinct families of DSPs. Perhaps the best known and most widely deployed is the CEVA-TeakLite DSP family, a classic memory-based architecture, with 16-bit or 32-bit word-widths and single or dual MACs. The CEVA-X DSP family offers a combination of VLIW and SIMD architectures, with different members of the family offering dual or quad 16-bit MACs. The CEVA-XC DSP family targets software-defined radio (SDR) modem designs and leverages a unique combination of VLIW and Vector architectures with 32 16-bit MACs.

Analog Devices produce the SHARC-based DSP and range in performance from 66 MHz/198 MFLOPS (million floating-point operations per second) to 400 MHz/2400 MFLOPS. Some models support multiple multipliers and ALUs, SIMD instructions and audio processing-specific components and peripherals. The Blackfin family of embedded digital signal processors combine the features of a DSP with those of a general use processor. As a result, these processors can run simple operating systems like μCLinux, velOSity and Nucleus RTOS while operating on real-time data.

New Words

latency	[ˈleɪtənsi]	*n.* 潜伏；潜在因素
defer	[dɪˈfɜː]	*v.* 使推迟，使延期；推迟，延期，服从
arbitrary	[ˈɑːbɪt (rə) ri]	*adj.* [数] 任意的；武断的；专制的

assembly [ə'sembli] *n.* 装配；集会，集合
compiler [kəm'paɪlə] *n.* 编译器；[计] 编译程序；编辑者，汇编者
matrix ['meɪtrɪks] *n.* [数] 矩阵；模型
convolution [ˌkɒnvə'luːʃ(ə)n] *n.* [数] 卷积；回旋；盘旋；卷绕
polynomial [ˌpɒlɪ'nəʊmɪəl] *n.* [数] 多项式；由 2 字以上组成的学名
superscalar ['suːpə, skeɪlə] *n.* 超标量体系结构
saturation [sætʃə'reɪʃ(ə)n] *n.* 饱和；色饱和度；浸透；磁化饱和
modulo ['mɒdjʊləʊ] *prep.* 以……为模 *adv.* 按模计算
peripheral [pə'rɪf(ə)r(ə)l] *adj.* 外围的；次要的 *n.* 外部设备
milestone ['maɪlstəʊn] *n.* 里程碑，划时代的事件
operand ['ɒpərænd] *n.* [计] 操作数；[计] 运算对象
leverage ['lev(ə)rɪdʒ] *n.* 手段；杠杆作用 *v.* 利用；举债经营

Notes

[1] PSTN (Public Switched Telephone Network)：公共交换电话网络，一种常用的旧式电话系统，即我们日常生活中的电话网。

[2] Texas Instruments (TI) 是一家位于美国得克萨斯州达拉斯的跨国公司，以开发、制造、销售半导体和计算机技术闻名于世，主要从事数字信号处理与模拟电路方面的研究、制造和销售。它在 25 个国家有制造、设计或者销售机构。它是世界第三大半导体制造商，仅次于英特尔和三星；是移动电话的第二大芯片供应商，仅次于高通；同时也是在世界范围内的第一大数字信号处理器 (DSP) 和模拟半导体组件的制造商，其产品还包括计算器、微控制器以及多核处理器。

[3] SIMD (single instruction multiple data)，单指令多数据流，能够复制多个操作数，并把它们打包在大型寄存器中的一组指令集。

[4] VLIW (very long instruction word) 是一种非常长的指令组合，它把许多条指令连在一起，提高了运算的速度。VLIW 体系结构是美国 Multiflow 和 Cydrome 公司于 20 世纪 80 年代设计的体系结构，EPIC 体系结构就是从 VLIW 中衍生出来的。

Questions for Discussion

1. What is the advantage of a specialized digital signal processor?
2. Do DSPs support virtual memory? Why?
3. Can modern signal processors perform well? Why?

课文参考译文

第1单元

课文A 电信的历史

远程通信的历史从非洲、美洲及部分亚洲国家使用烟雾信号和鼓点开始。18世纪90年代，首个固定的旗语通信系统出现在欧洲；然而，直到19世纪30年代，电气通信系统才出现。本文细述了电信的历史和成就了今天的电信系统的人物。电信的历史是整个通信史中的重要部分。

1. 古老的系统和旗语

早期的远程通信包括烟雾信号和鼓点。非洲的本土居民、新几内亚人和南美人使用信息鼓[1]，北美和中国使用烟雾信号。与大家所想的不同，这些系统不仅仅用来宣示军营的存在。

中世纪时，在山顶上使用一串信号灯被普遍作为传递信号的方式。信号灯链的主要缺点是它们只能传递很少的信息。因此，如“发现敌情”这一消息的意义必须提前达成一致。关于使用信号灯的一个很典型的例子是在西班牙无敌舰队时期，当一连串的信号灯从普利茅斯传递到伦敦时则示意了西班牙战舰的到来。[2]

法国工程师克劳德·沙普在1790年开始研究视觉电报，通过利用多对“钟表”的摆臂构造出不同的符号。但这种方式在远距离时并不可行，于是沙普改进了他的模型，采用两组铰接的木梁。操控者用曲柄和绳子移动木梁。他在里尔和巴黎之间建了他的首个电报线路，随后是一条从斯特拉斯堡到巴黎的线路（见图1-1）。

然而，旗语作为一种通信系统，既需要熟练的操控者又要求每10～30km建立一个昂贵的塔台。因此，最后一条商业线路在1880年被废弃了。

2. 电报机

利用电进行通信的实验始于1726年，起初并未成功。参与的科学家包括拉普拉斯、安培和高斯。首个能够运行的电报机由弗朗西斯·罗纳兹于1816年制成，采用了静电原理。

查尔斯·惠特斯通和威廉·福瑟吉尔·库克发明了一项五针六线系统专利，并于1838年投入商业应用。它利用指针的偏转来表达信息，并在1839年4月9日开始运用于21km长的大西部铁路上。查尔斯和库克二人将他们的装置视为“现存的电磁电报的改进”，而不是一项全新的设备。

在大西洋的另一边，塞缪尔·莫尔斯研制出了一款他曾在1837年9月2日演示过的电报机。阿尔弗雷德·韦尔看到了该演示后加入了莫尔斯的队伍，并致力于寄存器的研发，这是一种集成信息记录装置和纸带的电报终端设备。该终端设备已获得成功的演示，从1838年1月6日超过3mile（5km）的距离到1844年5月24日最终超越了华盛顿与巴尔的摩之间40mile（64km）的距离。这项发明专利收益颇丰，到1851年美国的电报线跨越了2万mile

(3.2万km)。莫尔斯对于电报最重要的贡献是他与韦尔一同发明的简单并且高效的莫尔斯码，仅需要两根电线即可实现，这是在惠特斯通复杂且昂贵的系统之上一次巨大的进步。[3]莫尔斯码的通信效率超过了100年来数码通信的哈夫曼编码，但是莫尔斯和韦尔以纯粹经验为主发明了这个电码，用更短的编码代替较频繁的字母。

3. 电话

电话发明于19世纪70年代，它基于早期的谐波（多信号）电报。首个商业电话业务于1878年和1879年在大西洋两岸的纽黑文市和伦敦市建立。亚历山大·格雷厄姆·贝尔享有两国都需要的这种服务——电话的专利。所有其他的电话设备和产品的专利均起源于贝尔的这项主专利。电话的发明归属一直备受争议，关于这个话题不时地会有新的争议出现。正如其他伟大的发明，如收音机、电视、电灯、数字计算机，有几位发明家进行电线传输声音的先驱试验，然后在彼此的想法上不断改进。[4]然而，最主要的创新者是亚历山大·格雷厄姆·贝尔和伽德纳·格林·哈伯德，他们创建了第一家电话公司——美国的贝尔电话公司，后来演变成为美国电话电报公司（AT&T），这家公司是世界上最大的电话公司。

首个商业电话业务于1878年和1879年在大西洋两岸的康涅迪格州纽黑文市和英格兰伦敦市建立，此后这项业务便飞速发展。到19世纪80年代中期，随着市际线路的建立，美国的主要城市之间都实现了电话通信。1915年1月25日，首个跨州电话实现。尽管如此，跨大西洋的客户间语音交流直到1927年1月7日通过无线电波才得以建立。

4. 收音机和电视

1894年后的几年，意大利发明家古列尔莫·马可尼在空间电磁波（无线电传输）的基础上建造了首个完整、成功商业化的无线电报系统。1901年12月，马可尼在加拿大纽芬兰圣约翰市和英格兰康沃尔之间建立了无线电，为他赢得了一项诺贝尔物理学奖（他与卡尔·布劳恩共同获得）。1900年，雷金纳德·范森顿采用无线电传输了语音。

20世纪的大部分时间，电视采用的都是布劳恩发明的阴极射线管。这种电视的首个有成功的希望的版本是由菲洛·法恩斯沃思设计的，他于1927年9月7日向他在爱达荷州的家人展示了粗糙的轮廓图像。法恩斯沃斯的设备可与卡门·提安逸和弗拉基米尔·兹沃雷金的协同工作匹敌。尽管这项设备的使用并不像所有人都期望的那样，它还是为法恩斯沃斯挣得了一家小型生产公司。1934年，他在费城的富兰克林研究所第一次公开展示了他的电视并开设了他自己的广播电台。

20世纪中后叶，同轴电缆和微波无线电中继器的普及使得电视网络能够扩展到更大的地区。

电视不只是一项仅限于其基本的和实际应用的技术。它还有器具的功能，能够讲述社会中发生的故事和传播信息。它是一个文化的工具，能够提供接受信息和体验想象的大众经历。它扮演着“世界之窗”的作用，它通过编排个人经历之外的故事、胜利、悲剧节目将全世界的观众连接在一起。

5. 卫星

首个中继通信的美国卫星是1958年的SCORE项目，它利用一个磁带录音机来存储并转发声音信息。它被用来向全世界传递美国总统德怀特·艾森豪威尔的圣诞问候。

电星1号是首个主动、直接转播商业通信的卫星。它属于美国电话电信公司，是美国电话电信公司、贝尔电话实验室、美国航空航天局、英国邮政总局、法国国家邮政局多国协商发展卫星通信的一部分。该卫星于1962年7月10日由美国航空航天局在卡纳维拉尔角发射，这是首次私人资助的航天发射。中继1号卫星于1962年12月13日发射，成为1963年11月22日首个跨太平洋播报的卫星。

首个也是历史上最重要的通信卫星应用是洲际长途电话。固定的公用交换电话网通过地面线路向地球站转播电话，并在那里通过地球轨道同步卫星传送给接收卫星的天线。通过使用光纤，海底通信电缆得到了改善，在20世纪晚期卫星拨号的使用减少了，但是它们依然是服务远程岛屿的唯一方式。例如，阿森松岛、圣赫勒拿岛、迭戈加西亚岛和复活节岛，这些岛屿没有海底电缆可用。

6. 计算机网络与互联网

1940年9月11日，乔治·斯蒂比兹利用电传打字机向他在纽约的复数计算器传送问题，并在新罕布什尔州的达特茅斯学院接受计算返回的结果。这种具有远程非智能终端的集中式计算机或大型机的配置在20世纪50年代一直很流行。然而，直到20世纪60年代，研究者们才开始探究分组交换技术——一项可以使多个数据块传送给不同的计算机，而无须预先传给一个集中主机的技术。1969年12月5日，一个四节点网络出现在加州大学洛杉矶分校、斯坦福研究院、犹他大学和加州大学圣巴巴拉分校之间。这个网络后来发展成了阿帕网络，到1981年由213个节点组成。1973年6月，首个非美国节点被加到属于挪威NORSAR项目的网络中，紧随其后的是伦敦的节点。

阿帕网络的发展集中于公开征求意见的过程，1969年4月7日《RFC1》公布。这个过程非常重要，因为阿帕网络最终会与其他网络合并形成互联网，并且现今互联网依据的协议也是通过该过程详细规定的。1981年9月，《RFC791》出台了互联网协议v4（IPv4），《RFC793》发布了传输控制协议（TCP）——因此创建了TCP/IP，也就是今天大部分互联网所依存的协议。与TCP不同，一个被称为用户数据报协议（UDP）的更宽松的传输协议于1980年8月28日被递交，并成为《RFC768》，但该协议不能保证数据块的有序传送。简单邮件传输协议，即SMTP，于1982年8月由《RFC821》公布。*http：//1.0*是1996年5月由《RFC1945》发布的一项超链接互联网应用的协议。

在20世纪后期，利用过去的电话和电视网络，互联网接入变得普及起来。

课文B　海因里希·赫兹传记

海因里希·赫兹[1]在一系列精彩的实验中发现了无线电波，证实詹姆斯·克拉克·麦克斯韦的电磁学理论是正确的。赫兹也发现了光电效应[2]，给量子世界的存在提供了一个重要的线索。频率的单位赫兹是以其名字命名的。

1. 学校

海因里希6岁时开始在汉堡的维查德·兰格博士学校学习。这是一所由著名教育家弗里德里希·维查德·兰格经营的男子私立学校。学校教育不受宗教影响。考虑到学生们的个体差异，学校采用以儿童为中心的教学方法。这所学校也很严格，它期望学生努力学习、互相竞争，成为班上的优等生。海因里希享受他在学校的时光，他的确在班上名列前茅。不一样的是，兰格博士学校不教大学入学需要的希腊语和拉丁语的经典名著。年幼的海因里希告诉父母，他想成为一名工程师。他们给他找学校时，认为兰格博士学校是最佳的选择，因为其可供选择的课程中包含了自然科学。

海因里希的母亲对其教育特别上心。她认为他有制造物品和画画的天赋，于是每周日在工学院给他安排制图术课程。当时海因里希 11 岁。

2. 家庭教育和构建科学仪器

海因里希 15 岁时离开了兰格博士学校，在家接受教育。他认为他终究要上大学。那时他接受了希腊语和拉丁语的辅导，为考试做好准备。

他擅长语言，这似乎是从父亲那里继承的天赋。给海因里希教过几堂阿拉伯语课程的语言专家雷德斯洛布教授告诉他父亲，海因里希应该学习东方语言。他以前从未遇到过如此有天赋的人。

海因里希也再次在家庭教师的帮助下，开始在家学习科学和数学。他非常喜欢繁重的学习。他母亲说："他拿书本坐着时，没什么能打扰他或吸引他的注意力。"

虽然他离开了师范学校，但他仍继续在每周日的早上参加工学院的课程。他晚上用双手劳作。他学会了操作车床。他建立模型，然后开始构建越来越多的复杂的科学仪器，比如分光镜。他使用这一仪器做自己的物理和化学实验。

3. 成为一名科学家

(1) 在慕尼黑大学学物理学

1876 年春天，他 19 岁，再次搬到德累斯顿学习工程学。仅几个月后，他被征召入伍，服了一年义务兵役。服完兵役后，20 岁的赫兹于 1877 年 10 月搬到慕尼黑，开始学习工程师课程。一个月后，在内心极度痛苦的情况下，他退出该课程的学习。他认为他想成为物理学家的想法高于一切。于是他就读于慕尼黑大学，选择了高等数学、力学、实验物理学和实验化学。在慕尼黑大学一帆风顺的一年之后，他转到柏林大学，因为那里有比慕尼黑更好的物理实验室。

(2) 柏林、亥姆霍兹和赞誉

21 岁的赫兹开始在柏林伟大的物理学家赫尔曼 · 冯 · 亥姆霍兹的实验室工作。亥姆霍兹已经意识到了赫兹的罕见天赋，于是立即让他从事某个问题的研究工作，因为亥姆霍兹对这个问题的解决方案非常感兴趣。该问题是亥姆霍兹和另一个名叫威廉 · 韦伯的物理学家之间激烈争论的主题。

在亥姆霍兹的鼓励下，柏林大学哲学系给任何能解决以下问题的人提供奖励：电流会随着惯性运动吗？或者，我们能够把问题设计成这种形式：电流有质量吗？或者，如赫兹构造的：电流有动能吗？

赫兹开始处理问题，并很快地沉浸在愉悦的日常工作中。他每天早上听分析力学或电磁学的讲座，在实验室做实验直到下午 4 点，然后在晚上阅读、计算和思考。

他亲自设计了他认为能回答亥姆霍兹问题的实验。他开始真正欣赏自己，并写道：

"我无法告诉你，让我直接从自然界为自己和别人获得知识，而不仅仅从别人和自己身上学习，这有多么令人满意。"

4. 获奖

1879 年 8 月，22 岁的赫兹获得金质奖章。在他展示的关于电流是否有质量的一系列高灵敏度的实验中，电流的质量小得令人难以置信。我们必须记住，在赫兹进行这项工作时，电子——电流的载体——甚至还没有被发现。1897 年，即赫兹工作后的 18 年后，约瑟夫 · 汤姆孙才发现电子。

其他物理学家开始注意到赫兹的成就非常耀眼——这位年轻的学生在物理学前沿进行了各种实验，根据需要亲自改造仪器。他晚上在家里发展的实践技能被证明是无价的。久负盛名的《物理学年鉴》杂志刊载了他的获奖作品。

亥姆霍兹察觉赫兹在实验方面的惊人才能，于是让他角逐柏林科学院提供的奖励：验证詹姆斯·克拉克·麦克斯韦的电磁学理论。麦克斯韦于1864年提出光是电磁波，并且其他类型的电磁波也是存在的。

5. 物理学博士

赫兹拒绝了该项目，他认为这个尝试要花上几年的时间，并且不能保证成功。他拥有雄心壮志，想尽快发表新的研究成果以确立声望。

他没有为奖励工作，但他进行了三个月技艺娴熟的电磁感应项目。他将其写成论文。1880年2月，23岁时，他的论文使他获得了物理学博士学位。亥姆霍兹很快任命他为助理教授。那年年末，赫兹写道：

“想不到的是，我越来越意识到我处于自我研究领域的中心；不管这是愚蠢，还是明智，但这是一种非常愉悦的感觉。”

赫兹待在亥姆霍兹的实验室直至1883年，在此期间，他在各种学术期刊上发表了15篇论文。

6. 基尔的数学物理学

赫兹是一位有才华的实验物理学家，但在柏林谋得讲师职务的竞争很激烈。然而，在亥姆霍兹的支持下，赫兹成为基尔大学的数学物理讲师。该职位要求的是理论能力而非实验能力，因此拓展了他的能力。他开始在基尔专攻麦克斯韦方程式，1884年5月，他在日记中写道：

“晚上钻研麦克斯韦的电磁学。只有电磁学。”

最后，赫兹写出了受人尊重的论文，文中比较了麦克斯韦电磁学理论与其他理论。他得出结论，麦克斯韦的理论看起来最有前途。事实上，他把麦克斯韦方程式改写为一种更简便的形式。

他后来写道：“起初，麦克斯韦理论是最优美的……麦克斯韦理论的基本假设与通常观点矛盾，得不到决定性实验证据的支持。”（1884年5月日记。）

7. 无线电波的发现

（1）设备精良的实验室和进攻最伟大的问题

1885年3月，赫兹搬到卡尔斯鲁厄大学，不顾一切地回到实验物理学。他28岁时获得了正教授职位。确切地说，他获得了两个正教授职位，这是他的盛誉象征。他选择了卡尔斯鲁厄，因为这里有最好的实验室设施。

他一边思考着研究的走向，一边思绪飘向了亥姆霍兹6年前没能成功说服他进行的奖励工作：通过实验证明麦克斯韦理论。赫兹决定这一伟大的事业将成为他在卡尔斯鲁厄研究的重点。

（2）火花改变了一切

在几个月的实验尝试后，表面上牢不可破的壁垒开始摇摇欲坠，这些壁垒曾挫败了所有证明麦克斯韦理论的尝试。

它最初开始于1886年10月的一次偶然的观察中，当时赫兹正给学生们展示电火花。赫兹开始深思电火花和它们在电路中的影响，并开始了一系列的实验，以不同方式生成电火花。

他发现了惊人的现象。火花在它们跳跃的电线间产生有规律的电振动。这种振动每秒来回移动的频率比赫兹以前在电气工程中遇到的任何情况都高。

他知道振动由快速加速和减速的电荷组成。如果麦克斯韦的理论正确，这些电荷将辐射出像光一样穿过空气的电磁波。

8. 创造和检测无线电波

1886 年 11 月，赫兹搭建了如图 1-2 所示的装置。

振荡器。两端有两个直径为 30cm 中空的锌球。每个球体与铜线相连，铜线向中间聚拢，并留出电火花跳跃的间隙。

他使用高电压的交流电穿过中间的火花间隙，产生火花。

火花在铜线内产生猛烈的电流脉冲。这些脉冲在电线里反射，以每秒将近 1 亿次的速度来回波动。

正如麦克斯韦预计的那样，振荡的电荷生成电磁波即无线电波通过电线周围的空气传播。一些电波到达铜线回路 1.5m 远处，在里面产生浪涌式的电流。巨大的电流致使电火花能够穿过回路的间隙。

这是试验的胜利。赫兹已经能够创造和检测到无线电波。他通过空气把电能从一个装置传递到 1m 多远的另一个装置上，不需要接线。

“我认为我发现的无线电波没有任何实际效用。”海因里希·赫兹在 1890 年写道。

9. 走得更远

在接下来的三年多时间里，在一系列精彩的实验中，赫兹完全证实了麦克斯韦理论。毋庸置疑，他证明了他的仪器能够生成电磁波，证明他的电子振荡器发出的能量能被反射、折射，能够产生干扰模式，生成像光一样的驻波。

赫兹的实验证明无线电波和光波属于同一家族，如今我们称之为电磁波谱。然而，不可思议的是，赫兹并没领会到他创造的电磁波具有丰碑式的现实意义。

这是因为赫兹是最纯粹的科学家。他只对设计实验，诱使自然界向其吐露神秘的事情感兴趣。一旦达到此目的，他将继续前行，把实际应用留给其他人去开发。

赫兹 1886 年 11 月首次生成的电波很快改变了世界。

1896 年，古列尔莫·马可尼[3]申请了无线电通信专利。1901 年，他从英国发送无线信号越过大西洋到达加拿大。

赫兹的发现是现代通信技术的基石。收音机、电视机、卫星通信和移动电话都依靠它。甚至微波炉也都使用电磁波：电磁波能够穿透食物，从内部迅速加热。

我们检测无线电的能力已转向天文学。射电天文学使得我们能够“看到”我们在可视光谱中看不到的特征。因为闪电发出无线电波，我们甚至能听到木星和土星上的雷电风暴。科学家们和非科学家们都欠了海因里希·赫兹很多。

第 2 单元

课文 A　移动无线概述

当今网络互连的世界正在进行快速的技术转型。此转型的特点是电信基础设施的融合，用融合的 IP 数据网络提供集成的声音、视频和数据服务。

随着转型的进行，相应的通信标准也在持续发展。为了能够促进电信和IP网络的融合，调动网络的性能和提供新的多媒体服务，一些新的标准正在制定中。

1. 移动无线技术介绍

无线通信的相关技术很复杂，难以区分。无线技术已存在了一段时间。然而，新的无线标准近来有了快速的发展，用以支持声音、视频和数据通信的集成。这一快速发展或革命很大程度上由于人们追求无处不在的实时信息访问以及把网络融入个人商业活动所致。“忙忙碌碌”的人们希望互联网接入能随身移动，以便随时随地获取信息。

有很多可用于描绘无线技术特征的要素：

- 频谱或网络运行的频率范围；
- 所支持的传输速度；
- 主要的传输机制，如频分多址（FDMA）、时分多址（TDMA）或码分多址（CDMA）；
- 架构实现，例如以企业为基础的（建筑物内部的）、固定的或移动的架构。

此外，还可以通过很多要素区分移动无线技术，如全球移动通信系统（GSM）、TDMA、CDMA，包括以下方面：

- 传输功率控制；
- 无线电资源管理和信道分配；
- 编码算法；
- 网络拓扑和频率复用；
- 切换机制。

顾名思义，移动无线通信用以称呼那些支持用户流动性的无线技术，它能够提供无缝连接和不间断的实时业务。[1]移动无线技术支持网络访问，不论用户在无线覆盖的住所内还是住所外漫游。

2. 与蜂窝网络和移动无线相关的基本网络元素概述

本部分简短地介绍了与现有电信基础设施相关的一些基本网络组件，特别讨论了基于时分复用模式的无线网络基础设施的组件，该无线网络的部分组件最终将被基于IP的新的组件所替代。

20世纪80年代初期，移动无线通信采用了基于模拟技术的蜂窝网络进行支撑，如高级移动电话系统[2]。很多与蜂窝网络相关的技术实体仍在今天的无线网络中起着关键作用。随着无线通信技术的持续发展以及IP数据网络进一步融入现有的基础设施中，这些实体的一些功能可能仍存在于网络中，但会以不同的、更有效的方式执行。

以下网络要素是典型的蜂窝通信网络的一部分：

- 公用电话交换网（PSTN）；
- 移动交换中心（MSC）；
- 基站（BS）；
- 无线接入网（RAN）；
- 归属位置寄存器（HLR）；
- 访问位置寄存器（VLR）；
- 认证中心（AUC）。[3]

（1）公用电话交换网

公用电话交换网（PSTN）是基础，并保留着目前支持全世界成千上万用户联系的主要基础设施。公用电话交换网有几千英里的传输基础设施，包括固定的陆上线路、微波和卫星通信线路。20 世纪 80 年代早期和中期引进蜂窝电话系统后，随着移动无线通信服务的快速发展，公用电话交换网依然是固网的支撑，其采用 7 号信令系统（SS7）协议在分组交换环境中传输控制信息和信令消息。

（2）移动交换中心

移动交换中心（MSC）是移动无线网络基础设施的一部分，通常设立在移动电话交换局（MTSO），其提供以下服务：

• 如果电话是手机到路线的呼叫，则将语音业务从无线网络交换到公用电话交换网；如果电话是手机对手机呼叫，则将语音业务交换到无线网络的另一个移动交换中心。

• 提供电话交换服务和控制电话和数据系统之间的呼叫。

• 给网络提供机动功能，并可为多达 100 个基站子系统充当集线器。更具体地说，移动交换中心提供以下功能：

• 用户的移动管理（用户登记、用户服务认证和授权、登录网络，保留用户的临时位置信息，这样用户就能接收和发起语音电话）。

在全球移动通信系统中，移动交换中心的一些功能被分配给基站控制器（BSC）。在分时多址系统中，基站控制器和移动交换中心是集成的。

• 呼叫建立服务（基于被叫号码的呼叫路由）。这些电话可通过移动交换中心连接到另一个移动用户，或通过公用电话交换网连接到有线用户。

• 连接控制服务决定呼叫如何路由及如何建立中继来把承载的业务传递给另一个移动交换中心或公用电话交换网。

• 业务逻辑功能是指为用户路由电话所需的业务，比如 800 业务、呼叫转接或语音信箱。

• 代码转换功能，将语音话务从移动设备解压到公用电话交换网以及将语音话务从公用电话交换网压缩到移动设备。

（3）基站

基站（BS）是移动无线网络接入基础设施的组成部分，是空中接口的终点。通过空中接口，用户通信业务能够被来回地传送到移动台（MS）。在全球移动通信网络中，基站被称为基站收发台。

（4）无线接入网

无线接入网（RAN）可以看作无线网络的一部分，其处理无线电频率（RF）、包括信令系统的无线资源管理（RRM）以及通过空中接口传输数据的同步。

在全球移动通信系统网络中，无线接入网通常由基站收发台（BTSs）和基站控制器（BSCs）组成。用户会话从移动台连接到 BTS，再连接到 BSC 上。BTS 和 BSC 的组合称为基站子系统（BSS）[4]。

（5）归属位置寄存器

归属位置寄存器（HLR）是一个包含移动网络用户信息的数据库，该移动网络由专门的服务供应商维护。此外，关于漫游的用户，归属位置寄存器可能包含来访用户的服务配置文件。

移动交换中心使用归属位置寄存器提供的用户信息来认证和登记用户。归属位置寄存器存储“长期的”用户信息（而不是访问者位置寄存器管理的临时用户数据），包括移动用户的服务配置文件、位置信息和活动状态。

(6) 访问位置寄存器

访问位置寄存器（VLR）是移动交换中心维护的数据库，存储漫游到移动交换中心覆盖范围内的临时用户信息。

访问位置寄存器通常是移动交换中心的一部分，与漫游用户的归属位置寄存器通信，请求数据并把用户当前的位置信息保留在网络上。

(7) 认证中心

认证中心（AUC）为服务提供商提供手机认证和加密服务。在今天大多数的无线网络中，认证中心配置有位置寄存器，并通常作为归属位置寄存器复合体部分来使用。

3. IP 融入移动无线的模式

IP 数据网络和现有的电信基础设施的整合标准正在迅速地发展，并开始在今天的生产网络中实现。

图 2-1 表明以当代产业方向为基础的 IP 集成模式，反映了移动无线互联网论坛（MWIF）的一些最新的想法。MWIF 是服务供应商和设备供货商的预标准联盟，为了能在基于 IP 的移动无线网络上实现合作。MWIF 影响了诸如第三代合作伙伴计划（3GPP）和第三代合作伙伴计划 2（3GPP2）之类的标准体系，并成功采用了新的模式。

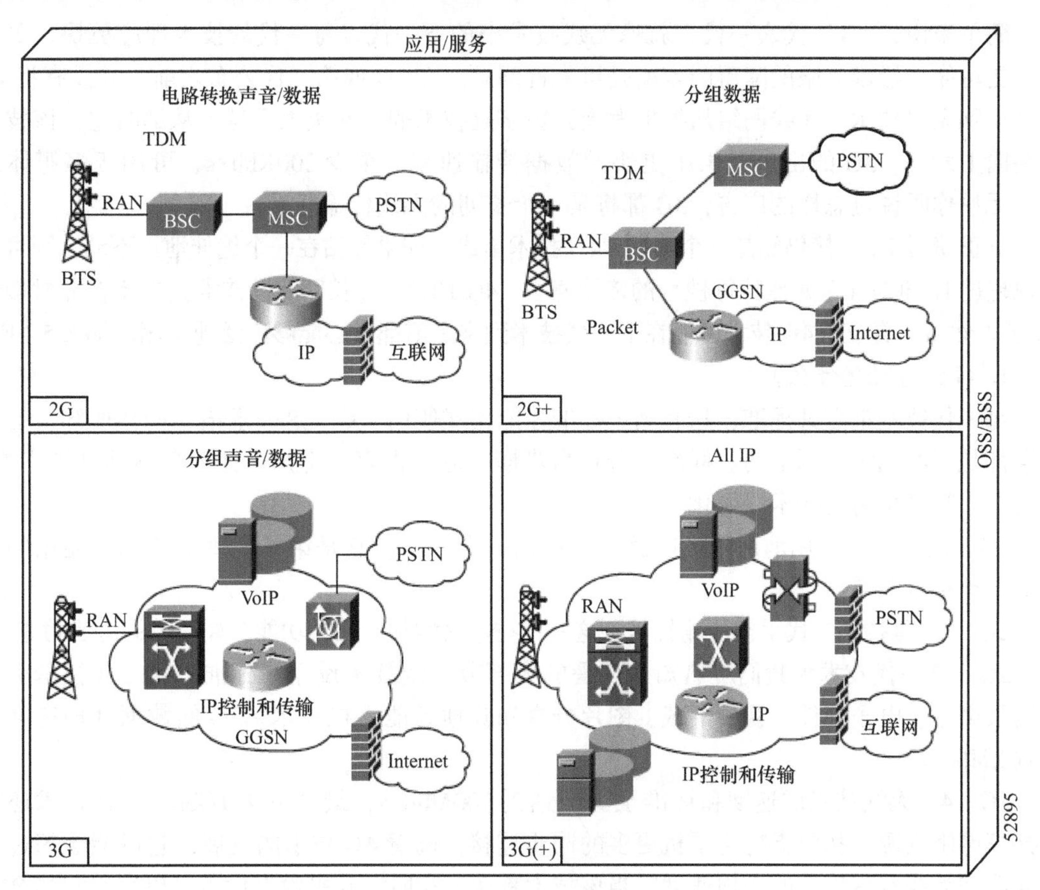

图 2-1 移动无线 IP 集成阶段

特别是，图 2-1 所示表明了应用于全球移动通信网络的思科系统通用分组无线业务网关支持节点的产品适用于此模式。

图 2-1 上面的两个象限表明我们今天在通信行业和 IP 数据服务设施所处的位置。第一个象限表示基于电路交换的语音和数据业务的基础设施的第一阶段。开始使用思科系统 V. 110 解决方案构建语音和数据集成的核心 IP 传输。

第二个象限描述了 2G + 技术的实施阶段，例如通用分组无线业务，这些技术能支持较高的传输速率。在此象限中，思科系统的通用分组无线业务网关支持节点提供 IP 分组数据服务。它充当来自 GSM 移动环境的数据进入互联网、其他公众网络和私有网络的通信网关。此阶段预计的服务包括执行一直开启的数据服务，使运营商能够通过数据包收费，而不是通过连接时间收费。CDMA 无线网络的分组数据服务节点（PDSN）也支持类似的服务。

第三个象限表示 IP 网络集成的第三阶段，声音和数据从无线接入网或外部的无线网控制器（RNC）合并到基于数据包的基础设施中。这被认为是 3G 解决方案。第三阶段使声音和数据应用合并，降低了成本。此外，还分配了一些 MSC 的组件或功能。

第四个象限表示最后一个阶段，包括 3G 服务以及以 IP 为基础的无线应用和移动组件，从而发展了真正的端对端、全面的 IP 无线网络解决方案。

课文 B　2G、3G、4G、4G LTE、5G 分别是什么

简单来说，“G”代表一代，就像无线技术中的下一代。每一代的技术理应更快、更安全、更可靠。最难克服的障碍因素就是可靠性问题。在 2G 或第二代公布之前，并没有将 1G 用来识别无线技术。无线网络从模拟发展到数字是技术的一次重大飞越。从那时起，该技术一路向上发展。3G 的出现提供了更快的数据传输速率，至少 200Kbit/s，可用于多媒体应用。无论你听说过怎样的广告，3G 都将是一个长期的无线传输标准。

真正使用 4G 连接仍然是一个难题。4G 技术承诺，如果你站在一个绝佳地点不动，则可以实现超过 1Gbit/s（千兆比特每秒）的传输速率。4G LTE[1]（长期演进技术）差不多非常接近 4G 了。然而，真正的 4G 技术可能在下一代技术到来之前都未必能够广泛地使用。那么 5G 呢？

1. G 的标准是什么？

每一代技术都有其标准，只有满足标准才能正式使用“G”这一术语。如你所知，这些标准都是标准制定者制定的。标准本身相当难懂，但广告商一定知道如何巧妙地处理它们。在这里，我尽量将这些术语简化一点。

1G。1G 是 2G 可用前从未被广泛使用过的一个术语。它是第一代手机技术，能做的只是简单地接打电话。

2G。2G 是指第二代手机传输技术。这一代技术又增加了几个功能，如简单的短信功能。

3G。这一代技术为我们所熟知并热爱的大部分无线技术设立了标准。第三代技术引入了网页浏览、电子邮件、视频下载、图片分享和其他智能手机技术。3G 的数据处理速度可达约 2Mbit/s。

4G。4G 无线技术的速度和标准至少应达到 100Mbit/s，最高可达 1Gbit/s。该技术还需要共享网络资源，从而支持与手机更多的同步连接。随着 4G 技术的发展，它可能会超过家庭互联网无线宽带连接的平均速度。当该技术首次发布时，几乎没有设备可以容纳它的全部功能。真正的 4G 技术的覆盖范围仅限于都市圈。在覆盖区域之外，4G 手机会退回 3G 标

准。4G 首次投入使用时，只比 3G 快一点。4G 与 4G LTE 不完全相同，4G LTE 非常接近标准的要求。首次推出 4G 时，实际上主要无线网络运营商并没有欺骗任何人，它们只是将事实夸张了一点。一部 4G 手机肯定符合标准，但寻找真正符合标准的网络资源较为困难。也许你在网络能够承载真正的 4G 技术之前，就正在购买 4G 功能的设备了。你知道 4G 比 3G 快，所以你愿意为加速买单。当 5G 投入市场时，情况也可能是这样的。

4G。LTE。它是长期演进技术的缩写——LTE 听起来更好。这个时髦术语是 4G 的一个版本，该技术正迅速成为最新的广告宠儿，并且已经非常接近标准所设定的速度。当你开始听说高级 LTE 时，我们将讨论真正的第四代无线技术，因为这两种技术是当时国际电信联盟实施的仅有的两种真正的 4G 制式。但这些已经成为往事了，因为 5G 很快就要到来。接着就是 XLTE[2]，其带宽至少是 4G LTE 带宽的两倍，可用于任何有 AWS 频段的地方。Verizon（威瑞森）公司、T-mobile（德国电信）公司和 Sprint（斯普林特）公司拥有所有高级 LTE 技术，各运营商都加入了自己的无线技术来强化该频段。

5G。有传言称 5G 正在接受测试，但其实 5G 的标准还未正式公布。我们期待这一项新技术在 2020 年左右推出，但现今世界在快节奏的发展中，这一天可能会更快到来。看似遥远，实则很快，5G 的速度将达到 1 ~ 10Gbit/s。

2. 了解 4G 技术标准

术语“4G”提到了一个新的互联网连接速度标准，但这个词具体意味着什么呢？4G 是什么？

大多数用户都对“4G”标准很熟悉，因为大多数智能手机都使用这一通信标准。4G 在数据传输技术的演变中只是“第四代”的意思。第一代移动通信技术（1G）与模拟传输一同诞生于 1981 年。1992 年 2G 技术以数字信息交换的形式出现了。3G 技术于 2001 年亮相，包括了至少 200Kbit/s 峰值传输速率的多媒体支持。现在是真正的 4G 支持。毫无疑问，4G 的意思是“第四代”，也表明在 3G 技术上的一些进步。

3. 4G 标准是谁制定的？

4G 技术意味着要为移动设备提供所谓的“超宽带”接入，国际电信联盟无线电通信部门（ITU-R）创建了一套标准，网络必须满足该标准才能称得上 4G，即高级国际移动通信（IMT-Advanced）规范。

4. 4G 标准是什么？

首先，4G 网络必须基于所有互联网协议（IP）分组交换，而不是电路交换技术；使用正交频分多址（OFMDA）多载波传输法或其他频域均衡（FDE）法，而不使用现有的扩频无线电技术。此外，4G 网络的峰值传输速率对于快速移动网络的用户来说要接近 100Mbit/s，对于有本地无线接入或慢速移动连接的用户来说要达到 1Gbit/s。真正的 4G 技术也必须能在不同网络之间提供不丢失数据的流畅切换，并为下一代媒体提供高质量服务。

4G 技术的一个最重要特点就是消除了并行的电路交换模式，它的分组交换网络节点采用了网际网络通信协定第六版（IPv6）。目前使用的网际网络通信协定第四版（IPv4）标准，能够给设备分配的 IP 地址数量有限，这意味着必须使用网络地址转换（NAT）来创建可重复使用的地址。该解决方案仅仅是将问题屏蔽了，并没有最终解决它。IPv6 提供更多的可用地址，并将有助于为用户提供更合理的体验。

5. 了解 LTE 技术标准

随着寻求全球性的解决方案而非本地解决方案，无线数据传输市场正在快速地发展，全球性解决方案可彼此协作，能提供强大的数据传输速率，可预测通信量的高峰。在新的第四代或4G市场的占有率竞争中，发展最快的标准之一就是LTE技术，该技术承诺在当前手机和数据终端服务的基础上大幅改进。

LTE实际上是“长期演进技术”的意思，它的全名是3GPP LTE，3GPP是第三代合作伙伴项目的缩写。该项目制定了该技术的发布文件。通常情况下，运营商的无线或移动服务中采用了LTE技术，并作为4G技术进行市场推广。但由于该技术尚未满足4G的ITU-R要求，最好还是将LTE的标准看作“3.9G”。文中提到的4G的ITU-R要求包括网络上传和下载的最低速率，并定义了如何建立连接。LTE技术的新版本——高级LTE满足真正的4G网络要求，预计在明年内推出。

那么，LTE到底是什么呢？LET是基于GSM[3]/EDGE和UMTS/HSPA的网络技术，采用了新的调制技术提升了容量和速度。它的峰值下载速率为300Mbit/s，上传速率为75Mbit/s，传输延迟时间小于5ms。该技术还可以管理组播和播放流，处理快速移动电话。其演进分组核心（EPC）——基于IP的网络体系结构，可以将语音和数据无缝切换至使用GSM、UMTS或CDMA2000技术的旧模式手机发射塔。此外，该技术可使用1.4~20MHz的承载带宽，同时支持时分和频分双工通信。总体而言，LTE技术的新体系结构可降低营运成本，并且具有更强大的综合数据和语音能力。

6. 了解5G技术标准

5G是第五代无线技术，它会比4G更快。这一点显而易见，但能快多少是一个问题。现在，这个问题的答案还不那么明确，但5G技术的预期速率应该是1~10Gbps，相比4G标准的100Mbit/s到1Gbit/s的速率确实快了很多。但是，这一速率能否实现则是另一个问题，2020年左右我们就能知道结果了。2020年是5G技术的预计发布时间。我认为5G将远早于正式发布时间投入使用，对此我很有信心。早在达到真正的5G速率前，你就会听到这样的宣传——“我们的全新5G网络”“现在试试最新的5G手机吧”，就像推出4G技术时一样。

5G会有多快？图片比语言更加直观，这张来自移动智库（GSMA Intelligence）的图片就很能说明问题（见图2-2）。5G会很快。对你来说够快吗？对任何人、做任何事以及物联网（IoT）来说，5G应该都是足够快的。对5G的使用远远超越了你的智能手机和设备，它可以驱动你的汽车，允许汽车之间相互通信，并使连接中的所有事物受益。

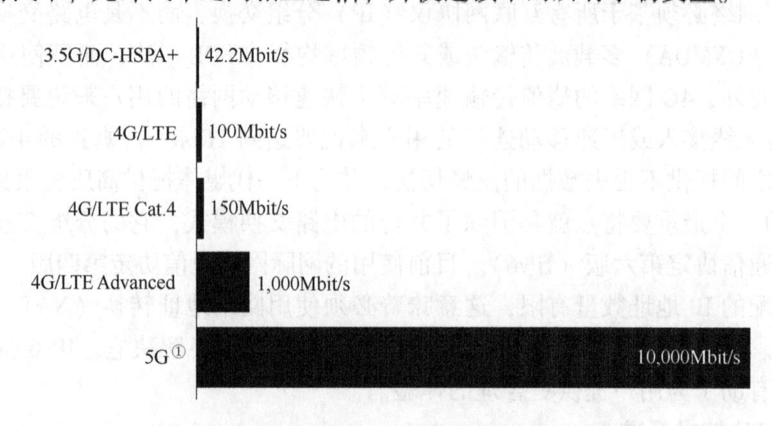

图2-2 各代技术的最高理论下行传输速率（Mbit/s）

①10Gbit/s为5G的最低理论上行传输限制速率。

第3单元

课文 A 互联网和通信

1. 什么是互联网

互联网已经彻底改变了我们的工作和娱乐方式。它使我们能够迅速交流、分享数据和搜寻信息。这一切都是因为计算机和网络的使用才得以实现。

互联网就是全球性的计算机网络。所有连接到互联网的计算机设备（包括个人计算机、笔记本式计算机、游戏机和智能手机）都是这个网络的一部分。总共有数十亿台计算机连接到互联网上，能够互相通信。如今，互联网已成为我们日常生活的重要组成部分（见图3-1）。

2. 互联网是如何起源的

在20世纪50年代，美国国防部以发展科技为目的成立了数个机构，如高级研究计划署（ARPA，现被称为DARPA）。然而，由于ARPA的科学家处在全国各地不同的大学里，所以他们很难交流或分享信息。为了解决这个问题，ARPA创建了一个名为阿帕网的计算机网络。在意识到阿帕网的巨大用处之后，其他机构也建立了自己的网络。但是，这些单个的网络之间难以相互通信。

20世纪70年代，一个被称为TCP/IP的协议发展起来。该协议允许不同的网络相互通信。这些单个网络连接起来便形成了一个庞大的广域网（WAN），这就是后来的互联网。

从那时起，组织和个人对互联网的使用逐年增长。起初，阿帕网仅由四台计算机构成。如今，已有数十亿台计算机连接到互联网。当我们连接到互联网时被称为“在线”。如今，互联网提供了许多在线服务，例如：

- 电子邮件和VoIP通信；
- 共享信息，如文字、图像、声音和视频；
- 信息存储；
- 流媒体电视节目、电影、视频、声音和音乐；
- 在线游戏；
- 购物；
- 社交网络；
- 银行业务。

这些在线服务大多是通过万维网上的网站来实现的。

3. 什么是万维网？

互联网是一个全球性的计算机网络。万维网是互联网的一部分，可以通过网站访问。网站是由网页组成的，而网页让你看到信息。

网站通过Web浏览器进行访问。浏览器是一个程序，用来显示网站上的信息。每个网站都有一个地址，通过该地址才能找到网站，有点像家庭地址。

网站的地址被称为网址（URL）。可以通过在Web浏览器中键入网址来访问网站。每个地址包含前缀“http:”，它告诉计算机使用超文本传输协议与网站进行通信。[1]然后浏览器会连接到互联网，找到该地址对应的网站，并将存储在该网站上的信息下载到计算机上以便我们查看。网站和网页是通过超链接连接到一起的。单击超链接，便可切换到另一个站点或页面。

4. 通过互联网传递信息

互联网是一个全球性的计算机网络，其中的一些计算机被称为网络服务器。网络服务器保存网站，为连接到互联网的其他计算机提供访问服务。存储网站被称为“托管”。网络服务器可以托管一个或多个网站及网页。向网站服务器发送信息称为上传，从网络服务器接收信息称为下载。

当你打电话时，你和你所呼叫的人之间形成了直接联系。当你正在通话时，其他人便不能和你交流。网络服务器需要能够同时与多台计算机进行通信。信息在上传或下载时，被分解成很多小块，称为数据包。每个数据包是客户端计算机和网络服务器之间的一次极其短暂的通信。因为每次通信只持续几毫秒，所以网络服务器看似可以同时与多台计算机通信。这有点像同时进行几个谈话，但依次只对每个人说一句话。

5. 使用 HTML 创建网站

互联网上的所有网页都是用一种被称为超文本标记语言（HTML）[2] 的语言创建的。HTML描述：

- 网页上显示的信息；
- 信息呈现在页面上的方式（格式）；
- 所有与其他网页或网站的链接。

HTML 可以通过专门的软件或简单的文本编辑器（如记事本）进行编写。只要文件以扩展名“. html”保存，它就能够被打开，并被浏览器视为一个网页。下面这段 HTML 代码示例了如何在网页上显示信息的（见图 3-2）：

```
<html>
        <body>
            <h1>Hello world</h1>
            <p>This is my first webpage</p>
        </body>
</html>
```

该代码使用标签来描述信息的外观：

- <html>说明该文档是 HTML 文档；
- <body>说明出现在页面正文中的信息；
- <h1>说明下列文本作为突出标题显示；
- <p>说明这是新段落的开始。

6. 使用电子邮件进行通信

人们使用互联网进行通信的两种主要方式是电子邮件和网络电话[3]。E-mail 是电子邮件的简称，相当于发送数字信件。每个电子邮件都有发送者、接收者和消息。它与手写信件最大的区别是，手写信件需要等待邮局交付，而电子邮件的发送和接收几乎是同时的（见图 3-3）。

电子邮件的利与弊

除了速度快外，电子邮件还有以下几个优点：

- 发送电子邮件不需要任何花费，而手写信件需要购买邮票。
- 同样的电子邮件可以同时发送给很多人。
- 文字、影片或声音剪辑等资源，可作为电子邮件的附件一同发送。
- 每封发送的电子邮件都有记录，因此可以回顾并查看发送的内容。
- 电子邮件送到时，收件人不必在场。电子邮件在深夜也可发送，收件人在第二天查看邮件时可以看到。这对电子邮件发送到世界的另一端有很大的好处。
- 电子邮件可以在一年的任何一天或任何时间发送。而邮局通常在周日或银行假日不进行派送。
- 电子邮件能够在各种互联网设备上发送和接收，如个人计算机、笔记本式计算机、游戏机、平板电脑或智能手机。
- 电子邮件可以自动转发到另一个网址上。

电子邮件也有缺点：

- 收件人只有在联网时才能收到电子邮件。
- 有时电子邮件的附件可能含有病毒。
- 垃圾邮件也是个问题。钓鱼邮件也是，它旨在诱骗人们泄露个人信息。
- 因为电子邮件可以传送到任何地方联网的数字设备上，所以人们难以离开它们。

7. 使用网络电话和视频会议

视频会议是在互联网上的实时视频流，因此人们无须在同一房间中便能面对面地交流。尽管最初视频会议主要用于商业界和学术界，但是现在全社会很多人都在使用。网络电话（VoIP）是使我们能够进行视频会议的技术。许多公司提供免费的 VoIP 服务，包括 Skype、苹果的 FaceTime 和谷歌的 Hangouts。

如果要与另一个人进行视频会议，双方都需要有连接到互联网的计算机设备，并配置监视器、摄像头、麦克风和扬声器。摄像头使监视器上看到的视频图像得以发送，麦克风使通过扬声器听到的声音得以传输。一个被称为 VoIP 客户端的程序负责处理通信。

许多笔记本式计算机、平板电脑和智能手机都配置视频会议技术，而且有很多 VoIP 应用程序可供使用，这些应用程序通常是免费的。

视频会议的优缺点

视频会议有以下几个优点：

- 能够看到对方，同时也能听到对方的声音。
- 可以向别人展示我们周围正在发生的事情。
- 节省时间，不需长途跋涉去与某人见面和交谈。如果对方在世界的另一端，受益会更大。
- 节省旅途的花费。
- 可以同时与不同地方的几个人进行视频会议。

视频会议也有其缺点：

- 每个想参加视频会议的人都必须有合适的硬件和软件。
- 用智能手机进行视频会议可能产生高额费用，因为数据使用量大。
- 交互视频流需要大量带宽。如果你的互联网连接质量不好，视频会议可能会非常费力。

课文 B 互联网通信类型

如果您使用互联网，那么您可能会使用互联网通信来联系家人、朋友或同事。从发送即时消息给朋友到发送电子邮件给同事，再到拨打电话、进行视频会议，互联网提供了多种沟通方式。

互联网通信的优点很多。由于您（或您的雇主）已经为互联网账户付款，您可以通过发送即时消息或拨打网络电话而不是标准的市话服务来节省电话费用。当然，所有技术都有缺点，互联网通信也有很多，如病毒、隐私问题和垃圾邮件。

像所有技术（特别是与互联网相关的技术）一样，我们在线交流的方式也在不断发展。在本周的《你知道吗?》一文中，我们将介绍一些最受欢迎的互联网通信形式。

1. 即时通信

互联网通信增长最快的形式之一是即时通信（IM）[1]。IM 可视为两人以上的基于文本的计算机会议。IM 通信服务使您能够与另一个人建立一种私人聊天室，以便通过互联网实时通信。通常，当您的好友或联系人列表中的某个人在线时，即时通信系统会提醒您。然后，您可以开启与该特定个人的聊天会话。

IM 变得如此受欢迎的一个原因是它的实时性。不像电子邮件，您将等待收件人检查他/她的电子邮件并发送回复，如果您想要访问的人在线，并且可以在您的 IM 联系人列表中显示，您的邮件将立即显示在他们屏幕的窗口中。

IM 不仅被数百万互联网用户用于联系家人和朋友，而且在职场中也越来越受欢迎。公司的员工可以立即访问不同办公室的管理人员和同事，并且能够在急需信息时而不必通过拨打电话来沟通。总的来说，IM 可以节省员工的时间，并帮助企业减少通信费用。

与 IM 相关的一些问题包括模拟程序和病毒传播。模拟程序是相当于垃圾邮件的 IM，它被从互联网上收集 IM 账户名称的机器人程序永久保存，这些机器人程序可以模拟人类用户通过即时消息将模拟程序发送到 IM 账户。模拟程序通常包含其制作者试图推销的一个网站链接。由于 IM 本身的性质，模拟程序比垃圾邮件更具侵扰性。这些广告和垃圾信息将在您的即时消息窗口弹出，您需要立即处理，而通过电子邮件，您通常可以过滤掉很多信息，并在以后处理。此外，病毒和木马可以通过 IM 渠道传播。当 IM 用户收到某个网站的链接消息并下载了其恶意代码时，这些恶意的程序就会传播。该消息仿佛来自已知的 IM 联系人，这就是为什么收件人更有可能单击超链接并下载该文件。使用安全聊天规则（例如从不点击链接）和保持系统中反病毒程序的更新将有助于减少通过即时消息传播的恶意程序感染的机会。

2. 互联网电话和 VoIP

互联网电话由硬件和软件综合构成，您可以使用互联网作为电话的传输媒介。对于那些能够免费或以固定价格接入的互联网用户来说，互联网电话软件本质上可以在世界任何地方提供免费电话。在最简单的形式中，个人计算机之间的互联网电话就像将麦克风连接到计算机一样简单，并通过有线调制解调器将语音发送给具有与您的互联网拨号软件兼容的人。然而，这种互联网电话的基本形式并非没有问题。连接通话线路比使用传统电话要慢，语音的传输质量也不会接近传统电话的质量。

VoIP 是另一种基于互联网的通信方式，正日益普及。VoIP 硬件和软件协同工作并使用互联网传输语音电话，其利用 IP 协议发送语音数据包，而不是传统的电路传输，这称为 PSTN（公共交换电话网络）。语音流量被转换成数据包，然后通过互联网或任何 IP 网络路由，就像正常的数据分组被传送一样。当数据包到达目的地时，它们将再次转换成语音数据给接收者。

就像为您的互联网连接找到互联网服务提供商（ISP）一样，您需要选择 VoIP 提供商。一些服务提供商可以提供套餐，其中包括对其网络上其他用户的免费呼叫和基于固定通话时长的其他 VoIP 呼叫的统一收费。当您使用 VoIP 拨打长途电话时，您极有可能会支付额外费用。虽然这听起来很像常规电话服务，从您不再需要为您的每月话单支付额外费用便可看出，但它比传统的语音通信便宜。

3. 电子邮件

E-mail 是电子邮件的缩写，它是通信网络上消息的传输。消息可以是从键盘输入的说明或存储在磁盘上的电子文件。大多数大型机、小型计算机和计算机网络都有一个电子邮件系统。一些电子邮件系统仅限于单个计算机系统或网络，但其他电子邮件系统则可以连接到其他计算机系统，使您能够在世界任何地方发送电子邮件。

使用电子邮件客户端（诸如 Microsoft Outlook 或 Eudora 之类的软件），只要您知道任何一个人的电子邮件地址，您就可以撰写电子邮件并将其发送给他。所有在线服务和互联网服务提供商提供电子邮件和支持网关，以便您可以与其他系统的用户交换电子邮件。通常，电子邮件到达目的地只需几秒钟。这是群组通信特别有效的方法，因为您可以一次向群组中的所有人广播消息或文档。

电子邮件所受到的最大影响之一是垃圾邮件。尽管定义有所不同，但垃圾邮件可以被认为是任何电子垃圾邮件（通常是一些产品的电子邮件广告），它即使没有发给数百万人，也发送给数千人。垃圾邮件经常用来传播木马和病毒。因此，使用最新的防病毒程序非常重要，它可以扫描传入和传出的电子邮件病毒。

4. 互联网中继聊天

IRC 是互联网中继聊天的简称，是一个多用户聊天系统，允许人们聚集于“频道”或“房间”进行群聊或私聊。IRC 基于客户端/服务器模型。也就是说，要加入 IRC 讨论，您需要一个 IRC 客户端（如 mIRC）和互联网接入。IRC 客户端是一个在您的计算机上运行并从 IRC 服务器发送和接收消息的程序。反过来，IRC 服务器负责确保将所有消息广播给参与讨论的每个人。多个讨论可以同时开展，每个都被分配给唯一的频道。一旦您加入 IRC 聊天室（聊天室讨论由主题指定），您可以在所有参与者都能看到的公共聊天室中键入消息，也可以向单个参与者发送私人消息。拥有许多 IRC 客户端，您可以轻松创建自己的聊天室，并邀请其他人加入您的频道。您还可以设置密码保护您的聊天室，与您邀请的人进行更多的私人讨论。

5. 视频会议

视频会议是指在不同地点的两个或多个参与者之间通过使用计算机网络传输音频和视频数据的会议。每个参与者都有一台摄像机、麦克风和扬声器连接在他或她的计算机上。当两个参与者交流时，他们的声音被输送到网络上并传递到对方的扬声器，而视频摄像机前面的任何图像都会出现在对方的显示器窗口中。

为了使视频会议正常运行，会议参与者必须使用相同的客户端或兼容的软件。许多免费软件和共享软件视频会议工具可以在线下载，大多数网络摄像机也附带视频会议软件。许多更新的视频会议软件包也可以与公共 IM 客户端集成，用于多点会议和协作。

近年来，视频会议已成为远程教育的一种流行形式，可以经济、高效地提供远程学习、演讲嘉宾和多校合作项目。许多人认为，视频会议提供了标准 IM 或电子邮件通信无法实现的可视化连接和交互。

6. 短信和无线通信

短信服务（SMS）是一种全球无线服务，能够在移动用户和诸如电子邮件、寻呼和语音邮件系统等外部系统之间传输字母数字消息。消息不能超过 160 个字母数字字符，并且不能包含图像或图形。消息一旦发送，它将被短消息服务中心（SMSC）接收，然后将其发送到相应的移动设备或系统。随着无线服务的发展，多媒体消息服务（MMS）出台，其提供了一种将包含文本、声音、图像和视频的消息发送到支持 MMS 的手机的方式。

无线设备（如手机和掌上电脑）的通信频繁变化。今天，您不仅可以使用无线设备打电话，还可以发送和接收电子邮件和 IM。虽然您可以免费使用电子邮件、IRC 或 IM，但如果您有互联网账户，您将最终向移动运营商支付费用，从而在无线设备上使用这些服务。

第4单元

课文 A　光纤是未来发展趋势的五大原因

网络上到处都有关于光纤的讨论，特别是关于谷歌的千兆宽带服务，许多人可能想知道为什么光纤如此重要。这的确是一个值得思考的问题，而且事实上，光纤有许多优势，客户以及网络架构师类的群体可以享受到这些优势。在探讨这些个人利益之前，我们有必要后退一步，先来看看传统金属线材和光纤的基本差异。

1. 金属线材

自美国建国后，在超过 150 年的数据传输中，金属线材几乎一直被应用于传送电信号，这是很了不起的。现今的电线都比往昔的电线精致很多，但它们的工作方式是相同的：电线的一端通电，电流通过电线传输到远端，并作为信号被接收。沿着传输线路，信号强度会逐渐降低，因为能量要经受一种被称作阻抗的电摩擦。阻抗会导致信号随着传输距离的增加而衰减，而且电线温度会升高，这会造成一些问题，后文将概括说明。

2. 光纤

光能是未来的发展方向。光纤采用了特殊的纤维，其能够进行纯光的传输。就像电线一样，数据传输始于光纤电缆的一端并一路传输到终端，在终端被接收和解码。

优势 1：信号衰减更少。光会随着传输距离衰减，事实上，当太阳光明亮地照在我们身上，则表明了一件事：太阳光的传播距离远大于恒定的电磁脉冲波的传播距离，因为光的衰减速度远小于电的衰减速度。这对宽带运营商和其客户来说，意味着光纤可为更多的住宅和商业贸易区提供服务。那些知道数字用户线路没落的人可能还会记得距离对其造成的限制；如果住户的家门口附近没有数字用户线路[1]网络节点，那么他们可能就不是那么幸运了。随着时间的推移，情况有所改善，但光纤自出现的第一天就解决了这个问题。

优势 2：后备架空更多。因数据传输会产生热量，金属线早已接近其物理性能极限。电线组上过大的负荷会导致这些电线熔化成细渣并失去效用，这正是对旧电线需求过大时发生的情况。光纤传输光，而不是传输电，在大多数情况下，光的热效应可以忽略不计。光纤系统内的光尤其如此，其强度远远弱于太阳发出的紫外线光。

优势3：升级更加容易。虽然数字用户线路和电缆的服务供应商逐渐能够系统地提升其网络性能，但是整个行事的费用一直很高。在通常情况下，由于热问题，整段的电线需要被挖出，然后替换。光纤可以使得网络节点和变电站间的距离更远且可留有更多的后备架空，这就使得网络载体的升级不再那么麻烦。这也就意味着更少的费用会被转嫁到消费者身上。一个很好的例子就是威瑞森的光纤服务（FiOS）[2]网络，自公开上市以来，其性能已有极大的提高，这是数字用户线路和电缆服务从未有过的。

优势4：光纤是绿色环保的。大家可能开始思考，在金属导线上通过电传送数据会造成浪费吗？如果答案是肯定的，那么你是正确的；通过金属导线传输数据所需要的能量是光信号传送所需的能量的数十倍。而且，需要额外的变电站和节点来保持远距离传输时信号的强度，这只会增加金属线路的窘境。相比之下，光纤看起来要好得多。此外，网络升级时，在光纤栅栏侧的浪费要少很多。电信和有线电视供应商会如何处理所有这些“老的”需要升级的网络节点？谁知道呢，但光纤需要替换和升级的变电站和节点更少。还有一点，事实上，电缆几乎永远不会变坏，那为什么光纤被视为任何形式的金属线路和电力的绿色替代品，原因是显而易见的。

优势5：嘘……秘密。大型数字用户线路和电缆服务提供商已经知道光纤是多么有用，而且他们的网络有很好的机会使用光纤，以便更加接近他们的客户的住宅和商业贸易区，但他们不想让大家知道。在许多情况下，数字用户线路和/或电缆服务提供商的光纤网络其实离他们服务的许多顾客一英里左右远，为什么？因为他们知道光纤是最具成本效益的解决方案，并且他们知道，将光纤安置在他们服务的住宅和商业贸易区附近后，就大有希望过渡到全光纤网络，而且更加容易。

3. 新型光的发现可能会改变光纤的未来

爱尔兰物理学家开辟了量子力学的一片新天地。他们发现了一种新形式的光，此举挑战了迄今为止我们对光的观察和研究的本质认识。这也是意味着光纤和数据通信未来的大事件。

直到现在，光速一直被视为一个固定的常量。从字面上来说，你对光的认识是基于普朗克常数和角动量的，这是一个基于衡量光束的普朗克数学方程的一个倍数。现在，三一学院的一个团队发现了一个角动量，即一束光只有普朗克常数的一半数值，这使得它有资格作为一种新型光。

约翰·多尼根教授对找出将这种新型光应用于日常生活的方法很感兴趣。一个最直接和最有效的方法就是应用于光纤通信。光纤是计算机数据的基础。通过玻璃或塑料线来传送闪光，使得通过单波道传送高达400Gbit/s的数据成为可能。有了这个新发现，数据传输速率可以高得多且数据传输会更加安全。

现在，关于角动量不确定的看法已流传了一段时间了。这支队伍最终成功创建了一个测试环境，以验证是否如此。为了达到此目的，他们回顾光研究的基础知识，并向物理学家汉弗莱·劳埃德和数学家威廉·罗文汉密尔顿借鉴了一些经验。在1830年，他们观察到了锥形折射。利用晶体，他们看到一缕光线是如何形成单锥或单束光的。因为相关照片，普朗克发现了这些光的背后隐藏的数学公式。现在，约翰·多尼根教授、保罗·伊斯特汉助理教授和他们的团队可能将这一发现深入研究到下一个级别。

此发现还处于起步阶段，但都柏林大学三一学院纳米技术中心的主任和斯特凡诺·桑维托教授已意识到这一发现的重要性，“这一发现是物理界和科学界的一次突破”。数据通信公司一定会来拜访都柏林三一学院，并想要将这个新技术应用于大众传媒的前沿，这只是个时间问题。

由梅雷迪思·普拉克撰写

课文 B　光纤到 x

光纤到 x（FTTX[1]）是使用光纤为最后里程的通信提供全部或部分本地环路的所有宽带网络架构的通用术语。因为光纤能够承载比铜线更多的数据，尤其是在长途通信时，所以建于 20 世纪的铜线电话网络正在被光纤替代。

FTTX 是对光纤部署的几种配置的概括，其分为两组：FTTP/FTTH/FTTB（光纤一直铺设到用户所在地/家庭/楼宇）和 FTTC/N（光纤铺设到机柜/节点，用铜线完成连接）。

1. 定义

电信行业区分了几种不同的 FTTX 配置。如今最广泛使用的术语是：

FTTP（光纤到用户所在地）：该术语或者用作 FTTH 和 FTTB 的综合术语，或者用作包括家庭和小型企业的光纤网络。

FTTH（光纤到家/户）：光纤到达居住空间的边界，如住房外墙上的一个盒子。无源光网络和点对点以太网是直接从运营商中心局通过 FTTH 网络提供三网合一业务的架构。

FTTB（光纤到楼、企业或地下室）：光纤到达建筑物的边界。例如，多住宅单元中的地下室，并通过类似于路边或电杆技术的可选方式最终连接到个人居住空间。

FTTO（光纤到办公室）：光纤连接从主机房/核心交换机安装到位于用户工作站或服务点的特殊小型交换机（称为 FTTO 交换机）。该小型交换机通过标准双绞线接插线为终端用户设备提供以太网服务。这些交换机可以分布于整个楼宇中，但从一个中心点进行管理。

FTTN/FTTLA（光纤到节点、街区或末级放大器）：光纤终止于距离客户驻地几英里的街道机柜中，最后用铜线连接。FTTN 通常是迈向全面 FTTH（光纤到户）的临时步骤，通常用于提供“高级”的三网合一电信业务。

FTTC/FTTK（光纤到路边/路缘、壁橱或橱柜）：这与 FTTN 非常相似，但是街道机柜或电杆更靠近用户的场所，通常在 1000ft（300m）内，在该范围内可以采用诸如太网或 IEEE[2] 1901 电力线网络的高宽带有线技术和无线 WiFi 技术。FTTC 有时被含糊地称为 FTTP（光纤到电杆），会导致与不同的光纤到本地系统的混淆。

为了促进一致性，特别是在比较国家之间的 FTTH 渗透率时，欧洲、北美和亚太地区的三个 FTTH 技术委员会同意 2006 年 FTTH 和 FTTB 的定义，并于 2009 年、2011 年和 2015 年进行了更新。FTTH 技术委员会对 FTTC 和 FTTN 没有正式的定义。

图 4-1 所示说明了 FTTX 体系结构随着光纤与终端用户之间的距离而变化的情况。左边的建筑是中心局，右边的建筑物是中心局所服务的楼宇之一。虚线矩形表示同一楼宇内的独立居所或办公空间。

2. 优点

光缆可以长距离地高速传输数据，但传统电话线和 ADSL 中使用的铜线电缆却不能。例如，千兆以太网（1Gbit/s）的常见形式是运行在比较经济的 5e 类、6 类或增强型 6 类非屏

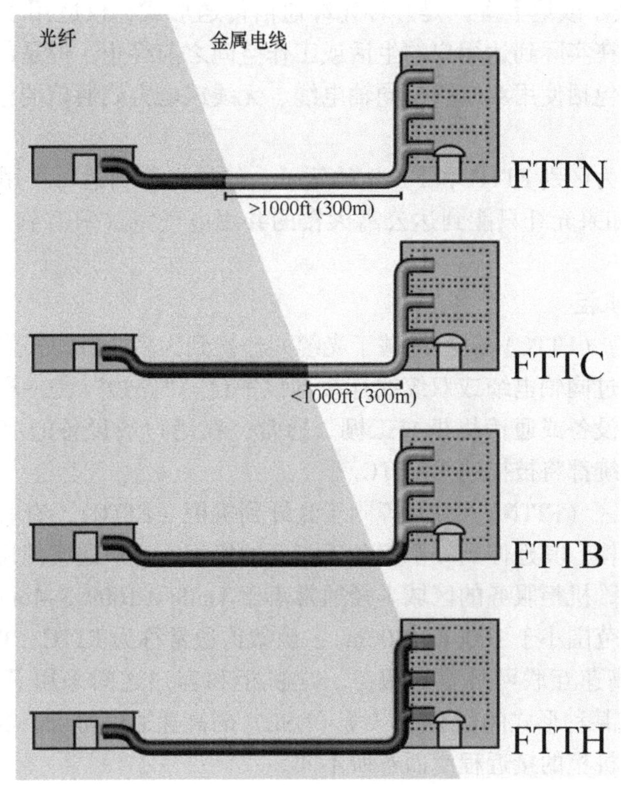

图 4-1

蔽双绞线上，但只能达到100m的距离。然而，光纤上的1Gbit/s以太网可以轻松地延伸到数十公里。因此，FTTP已经被世界上所有的主要通信供应商选中，其可以长时间地在1Gbit/s的对称连接上将数据直接传送到消费者家中。将光纤直接引入楼宇的FTTP配置可以提供最高的速率，因为剩余的距离可以使用标准的以太网或同轴电缆。谷歌光纤提供1Gbit/s的速度。

光纤通常被认为是“不会过时的技术”，因为连接的数据速率通常受到终端设备而不是光纤的限制，在光纤本身必须升级之前允许通过设备升级大幅度提高速度。尽管如此，所选择的应用光纤的类型和长度，如多模与单模，对将来超过1 Gbit/s连接的适应性至关重要。

FTTC（在街道机柜中光纤转变为铜线）通常离用户太远，因为标准以太网配置于现有的铜线电缆之上。它们通常使用超高速数字用户线路（VDSL），下行速率为80Mbit/s，但是在100m之外会快速下降。

3. 光线到用户所在地

FTTP（光纤到用户所在地）是一种光纤通信传送的形式，其中光纤在光分配网络中从中心局一直运行到用户所在地。术语“FTTP”已变得不怎么明确了，它也可能是指FTTC，即光纤终止于电杆而不到达用户住所。

可以根据光纤到达的位置对光纤到用户所在地进行分类：

FTTH（光纤到家）是一种光纤通信传输形式，可以到达个人生活或工作场所。光纤从中心局延伸到用户的生活或工作场所。在用户的生活或工作区域，信号可以通过任何方式传送到整个区域，包括双绞线、同轴电缆、无线、电力线通信或光纤。

FTTB（光纤到楼宇或地下室）是一种光纤通信传送形式，仅适用于包含多个生活或工作空间的楼宇。光纤在实际到达用户的生活或工作空间之前终止，但是却延伸到包含该生活或工作空间的楼宇，包括使用双绞线、同轴电缆、无线或电力线通信的任何非光纤方式将信号传送到最后距离。

公寓大楼可以作为区别 FTTH 和 FTTB 的例子。如果光纤到达每个用户公寓单元内的嵌板上，则为 FTTH。如果光纤只能到达公寓大楼的共用电气室（只有到底层或每层），就是 FTTB。

4. 光纤到路边/机柜

光纤到路边/机柜（FTTC）是一种基于光缆并连接到为多个客户提供服务的平台的电信系统。每个客户都通过同轴电缆或双绞线连接到该平台。“路边”是一种抽象，可以简单地理解为一个竿式安装设备或通信机柜或工棚。通常，在用户端设备的 1000ft（300m）范围内的任何终端光纤系统都将被描述为 FTTC。

光纤到节点或街区（FTTN），有时等同于光纤到机柜（FTTC），有时与之不同。这是一种基于光缆的电信架构，其延伸到一个服务于街区的机柜。客户通常使用传统的同轴电缆或双绞线连接到该机柜。机柜服务的区域半径通常小于 1mile（1609.344m），可容纳数百个客户。如果机柜的半径范围小于 1000ft（300m），该架构通常称为 FTTC/FTTK。

FTTN 允许提供高速互联网等宽带服务。在机柜和客户之间采用了诸如宽带电缆接入（通常为 DOCSIS）或某种形式的数字用户线（DSL）的高速通信协议。数据速率根据所使用的确切协议和客户对机柜的接近程度而有所不同。

与 FTTP 不同，FTTN 通常使用现有的同轴或双绞线基础设施来提供最后一公里的服务，因此部署成本较低。然而，从长远来看，相对于那些使光纤更贴近用户的应用，FTTN 带宽潜力是有限的。

对有线电视供应商而言，这种技术的一种变体被用于光纤同轴电缆混合（HFC）系统中。在到达客户（或客户附近）之前，该技术将取代模拟放大器直至最后一级，它有时会被缩写为 FTTLA（光纤到最后一级放大器）。

FTTC 允许提供宽带服务，如高速互联网。通常现有的线路与通信协议一起使用。例如，宽带电缆接入（典型的 DOCSIS）或连接路缘/机柜和客户的某种形式的 DSL。在这些协议中，数据速率根据所使用的确切协议而变化，并根据客户对机柜的接近程度而有所不同。

如果运行新的电缆是可行的，光纤和铜缆以太网能够圆满地以 100Mbit/s 或 1Gbit/s 的速度连接“路边”。即使使用相对便宜的户外第 5 类铜缆超过数千英尺，但所有以太网协议，包括以太网供电（PoE）都是被支持的。大多数固定无线技术依赖于 PoE，包括摩托罗拉 Canopy 系统，其具有能够在数百英尺电缆馈送的 12V 直流电源上运行的低功率无线电的能力。电力线网络部署也依赖于 FTTC。使用 IEEE P1901 协议（或其前身 HomePlug AV），现有的用户电缆从路缘/电杆/机柜到家庭每个交流电源插座可以提升到 1Gbit/s 的速率，覆盖范围相当于稳健的 WiFi 实施，并增加了单芯电缆的功率和数据的优势。

与新电缆及其成本和负担相比，FTTC 的部署成本更低。然而，从历史的角度来说，它比 FTTP 的带宽潜力更低。在实践中，光纤的相对优势取决于可用的回程带宽、基于使用的计费限制（这妨碍了充分利用最后一英里的性能）、客户端设备和维护限制以及运行光纤的成本（其可以随地形和建筑类型而变化）。

历史上，电话和有线电视公司都避免从入网点到客户端使用几种不同传输协议的混合网络。竞争成本压力的增加、三种不同的现有线路解决方案的有效性、智能电网部署要求（如查塔努加市）以及更好的混合网络工具（如阿尔卡特朗讯和高通创锐讯等主要供应商，边缘网络的 WiFi 解决方案、IEEE 1905 和 IEEE 802.21 协议以及 SNMP 改进）都使得 FTTC 更有可能地部署到经济不发达的地区，和 FTTP/FTTH 共同提供服务。

第 5 单元

课文 A　物联网

物联网[1]（程式化物联网或 IoT）是物理设备、车辆（也称“互联设备”和“智能设备”）、建筑以及其他物体的互联网络。这些物体嵌入了电子设备、软件、传感器、执行器和网络连接，使得它们能够收集和交换数据。2013 年，物联网全球标准行动举措工作组（IoT-GSI），将物联网定义为“信息社会的基础设施”。物联网允许在现有网络基础设施上远程感知和（或）控制物体，为物理世界更直接地集成到基于计算机的系统中创造了机会，从而使效率、精度和经济效益得到了提高，同时还减少了一定的人为干预。当物联网通过传感器和执行器扩张时，该技术就变成了更为普遍的信息物理系统的一个实例，其中也包括智能电网、智能家居、智能交通和智慧城市等技术。每一个物体在它所处的嵌入式计算系统中都是唯一可识别的，但在现有的互联网基础架构内能够进行交互操作。专家估计，到 2020 年物联网将包含近 500 亿个物体。

通常来说，物联网有望提供设备、系统和服务的高级连接，超越机器对机器（M2M）通信，并涵盖不同协议、领域和应用程序。预计会将这些嵌入式设备（包括智能物件）的互联引入几乎所有领域的自动化中，同时还能实现如智能电网等高级应用，并扩展到诸如智慧城市等领域。

在物联网意义上的“物”可指各种各样的设备，如心脏监测植入物、农场动物身上的生物芯片转发器、沿海水域里的电动蛤蜊、具有内置传感器的汽车、用于环境/食品/病菌监测的 DNA 分析装置，或协助消防员搜索和救援的野外作业设备。法律学者建议将“物”视为“硬件、软件、数据和服务的不可分混合体”。这些设备借助各种现有技术的帮助收集有用的数据，然后让这些数据在其他设备之间自动传输。目前的市场例证包括家庭自动化（也称为智能家居设备），如照明的控制和自动化，供暖设备（如智能温控器）、通风设备、空调（HVAC）系统以及使用 WiFi 远程监控的洗衣机/烘干机、机器人真空吸尘器、空气净化器、烤箱或冰箱/冰柜等家用电器。

除了将互联网连接的自动化扩展到众多新的应用领域之外，物联网预计还将从不同的地点生成大量数据，因此就必须实现数据快速汇集，还需要增加索引、存储，更有效率地处理这些数据。物联网是当今智慧城市[2]的平台之一，是智能能源管理系统。

物联网的概念和术语是由彼得 T. 刘易斯创造的，他于 1985 年 9 月在美国国会黑人同盟 15 届周末立法会议的美国联邦通信委员会（FCC）[3]会议上发表的演说中首次提出这一概念。

1. 应用

据 Gartner 公司（技术研究和咨询公司）称，到 2020 年物联网上将有近 208 亿件设备。

ABI咨询公司估计，到2020年将有超过300亿件设备以无线方式连接到物联网。根据2014年皮尤互联网研究项目组进行的一项调查研究显示，大多数予以应答的技术专家和互联网用户（83%）认同物联网/云这一概念，到2025年，嵌入式和穿戴式计算（和相应的动态系统）将产生广泛的影响和有益的效应。因此，很显然物联网将包含大量与互联网连接的设备。为了适应这一新型、新兴技术的发展，英国政府在2015年的预算中拨出了4000万英镑用于物联网的研究工作。前英国财政大臣乔治·奥斯本认为，物联网就是下一阶段的信息革命，它可以实现从城市交通到医疗器械到家用电器所有事物的互联。

与互联网集成就意味着设备将使用IP地址作为唯一的识别符。然而，由于网际网路通信协定第四版（IPv4）的地址空间有限（允许有43亿个唯一地址），物联网中的对象不得不使用IPv6才能适应所需的极大地址空间。物联网中的对象不仅是具有感测功能的设备，而且还可以提供驱动功能（如通过互联网控制的锁或灯泡）。在很大程度上，如果没有IPv6的支持，未来的物联网将不可能实现；因此，将来IPv6能否在全球广泛应用，对物联网未来的成功发展至关重要。

网络嵌入式设备具有有限的CPU、内存和电源资源，这种能力意味着物联网可应用于几乎所有领域。这种系统可以在各类场景中收集信息，从自然生态系统到建筑和工厂，因此可以应用于环境感测和城市规划领域。

另一方面，物联网系统不仅能够感测事物，还可以对事物的执行行为做出反应。例如，智能购物系统可以通过追踪用户的专用手机，来监测特定用户的购买习惯，然后向这些用户提供他们所喜爱商品的优惠信息，甚至是他们需要的物品的位置，放置物品的冷藏室会自动地将信息传到用户的手机上。其他有关感测和执行的实例反映在采暖、电力和能源管理以及巡航辅助运输系统的应用中。物联网的其他应用还可以扩展到家庭安防和家庭自动化方面。“生物互联网”概念的提出是为了描述生物传感器网络，它可以让用户使用云计算分析来研究DNA或其他分子。

然而，物联网的应用并不只限于这些领域。物联网的其他专业化使用情况也可能存在。本文将对一些最重要的应用领域进行概述。根据应用领域的不同，物联网产品可大致分为5类：智能穿戴、智能家居、智慧城市、智能环境和智能企业。这些市场中的物联网产品和解决方案的特点有所不同。

2. 实现物联网技术

有许多技术可以实现物联网。该领域的关键就在于物联网安装的设备之间通信所使用的网络，有好几种无线或有线技术都能满足要求。

（1）短程无线

- 低功耗蓝牙（BLE）——它提供的功率很小，有别于经典蓝牙，但通信范围不亚于经典蓝牙。
- 可见光无线连接（LiFi）——类似于WiFi标准的无线通信技术，但在增加的带宽上采用可见光通信。
- 近场通信（NFC）——可使两个电子设备在4cm范围内进行通信的通信协议。
- 二维码和条形码——机器可读的光学标签，可存储它所附着的物品的信息。
- 射频识别（RFID）——该技术使用电磁场来读取附在其他物品上的标签中所存储的数据。
- 线程（Thread）——基于IEEE 802.15.4标准的网络协议，类似于ZigBee，提供网际网路通信协定第六版（IPv6）寻址。

- 无线上网技术（WiFi）——广泛使用的基于 IEEE 802.11 标准的局域网技术，设备可以通过共享的接入点进行通信。
- WiFi 直连——点对点通信的 WiFi 标准的变体，无须接入点。
- 无线组网规格（Z-Wave）——提供短程、低时延的数据传输率以及低于 WiFi 功耗的通信协议，主要用于家庭自动化。
- 无线个域网（ZigBee）——基于 IEEE802.15.4 标准的个人区域网络通信协议，特点是低功耗、低传输率、低成本和高吞吐量。

（2）中程无线

- HaLow——WiFi 标准的变体，特点是在低传输速率下可以扩大低功耗通信的范围。
- 高级 LTE——用于移动网络的高速通信协议，规定了具有扩展覆盖范围、更高吞吐量和更低延迟的 LTE 增强标准。

（3）远程无线

- 低功耗广域网络（LPWAN）——用于在低传输速率下进行远程通信的无线网络，可降低传输的功耗和成本。
- 甚小孔径终端（VSAT）——使用小型蝶形天线进行窄带和宽带数据传输的卫星通信技术。

3. 有线

- 以太网——使用双绞线和光纤链路连接集线器或交换机的通用网络标准。
- 同轴电缆多媒体联盟（MoCA）——在现有的同轴电缆的基础上，实现高清视频和内容在整个家庭中分配的规范。
- 电力线通信（PLC）——利用电线传输功率和数据的通信技术。例如，家庭插电联盟规定使用 PLC 作为物联网的联网设备。

课文 B　忘掉物联网：“车联网”来了

如果大多数人能够在车辆共享的基础上做更多事情将会怎样？例如，用随需应变的车辆资源来代替传统的汽车所有权：夏天有敞篷车，冬天滑雪旅行有越野车。

如果驾驶技巧能够估算成某个分数，那么就可以实时提醒我们当心附近路上的糟糕驾驶人，并且能使导航系统提供更安全的可选路线时会怎样？设想一下，如果我们的世界里没有交通堵塞和事故将会怎样？或者如果我们的汽车能自动拾取我们的行李——并且犹如独立的豪华轿车服务一样把我们送到机场又将会怎样？

如果汽车制造商能通过与电信公司和其他想利用联网驾驶人终身收入机会的公司合作来资助我们购买汽车[1]的话会怎样？再考虑一下，保险供应商索取更高的保险费用（对于自己开车的人来说），或者让当地政府来监控个人 CO_2 排放量（作为免税或公路费用的交换）会怎样？

无论你是否接受这样的场景，它们都离我们不远了。这不仅是一次技术可行、联网车辆的进化，还远远超越了自驾车的范畴。并且，这不只是一个简单的传感器网络，它还是智能移动的时代——车联网时代。

基本上来说，汽车已经成为“终极移动设备”了，并且我们人类正在成为“联网驾驶人”。这些不是时髦术语：作为一个长期的战略建议者和空间分析者，我从 1998 年就用这些词语来描述汽车的根本性转变了，并且一直用到现在。例如，到2016年，在成熟的汽车市场（美国和西欧）中，大多数买家在购车时都会把汽车联网获取信息的能力作为一项关键的考虑指标。对于高端汽车品牌的买家，这个临界点很快在2014年达到，仅仅还有一年时间。

这种联网的汽车正在引领汽车产业进入自汽车问世之后的最重要的革新阶段。

1. 智能移动的年代将会改变一切

但它是什么呢？“联网汽车”是指在商业、人群、组织、基础设施和物体之间访问、消费、创建、充实、管理并且分享信息的汽车。这些物体包括其他车辆，这也就是为什么物联网变成了车联网。

随着这些车辆越来越多地联系起来，它们变得能自我感知、判断环境，并且最终能自主驾驶。这篇文章的读者或许能体验到你们人生中的自驾车——尽管不是汽车发展的全部三个阶段：自动驾驶到自主驾驶再到无人驾驶。

为了开发成功的汽车产品，我们仍需要解决许多技术、工程、法律和市场问题。但是这个汽车时代是建立在当前和相关产业发展趋势之上的。例如，数码生活方式的趋同，新的流动性解决方案的出现，人口结构的变化，以及智能手机和移动互联网的崛起。

消费者现在希望无论在哪里都能获得与之相关的信息……包括在汽车里。同时，这些技术正在形成一种更普遍和更有吸引力的新的汽车流动性解决方案——例如对等汽车共享。因为城市的汽车产权很昂贵，消费者特别是年轻人没有表现出和老一辈对汽车所有权同样的热情，这一点尤其重要。

为了获得成功，联网的汽车将会利用传感器、显示器、车内车外计算、车载操控系统、无线车载数据通信、机器学习、分析学、声音识别和内容管理等领先技术。这些将会带来可观的利益和机会：减少了事故率，提高了生产效率，改善了交通流量，减少了排放，扩大了电动车[2]的效用，新的娱乐选择，新市场和商业经验。

除了向汽车和驾驶人提供新功能，联网汽车还会扩大汽车商业模式来覆盖更广泛的产业——IT、零售、金融、服务业、传媒和家用电子产品。这一点很重要，因为它能挑战传统的汽车商业模式：企业不是仅专注于汽车的销售和维修，而是要关注汽车所代表的全部商业机会。

2. 但是消费者想要什么呢？

人们实际上都是这样想的吗？或者这只是商业思想家、技术人员和早期尝试者在回音室里预测的一个案例吗？答案是否定的。消费者确实对联网汽车的特点表现出了强烈的兴趣。例如，从高德纳公司2013年所做的分析中，我们发现所有美国的车主：

- 几乎一半（46%）的车主都对在车内稳定地访问移动应用感兴趣。这些应用包括接收所需的无线地图或软件更新，找寻可用的停车点和进行本地搜索；几乎40%的车主还选择了当零部件需要更换时能提醒他们的远程诊断功能。
- 超过1/3的人对自动驾驶、无人驾驶的汽车感兴趣。
- 30%的人可能选择允许他们的手机在车上联网的汽车。

越来越数字化的生活方式也会促使消费者去重新评估个人交通工具的选择。例如，将一个月的手机话费和住宅网络费用的综合成本与在加油站给车加油的成本进行比较。

与年龄大一点的车主（54岁以上）相比，这些权衡对于年轻一点的车主（18～24岁）来说更加重要。与仅为12%的老年人群相比，年轻人群（30%）更愿意选择联网车辆，但是两者具有相同比例的人数愿意使用车辆共享服务来代替买车。

显然，联网车辆的应用必须是安全、可靠和不分心的，要让消费者感性上称赞，理性上信服。从其他移动设备上简单地复制接口是远远不够的——车内的按钮实际上有很重要的特定功能。汽车产业需要引入新的体验并且彻底整合系统，这样消费者才不会觉得他们只是把 iPad 放在了乘客座椅上。

但事实是，汽车仍然停留在现状，将来才会被连接起来。这里描述的革新和改变会在未来的 20 年中相对快速地成熟起来。例如，我估计到 2016 年至少有三家公司会宣布具体的方案，因为即将到来的新产品发布会会提出无人驾驶车辆技术。

这并不是空想——只要细想一下汽车连接领域近来的一些进展：阿维斯出租汽车公司获得了“汽车共享”[3]，首个无线汽车软件被特斯拉解决；英特尔在联网汽车价值链中占重要地位；像斯普林特这样大的电信公司也将它们的领域发展到汽车行业中；一个苹果公司的高级主管进入了汽车制造商董事会。所有这些变化都标志着趋势所向。

对于那些也对汽车和驾驶充满热情的人来说，车联网的时代将为他们展现一个迷人的新世界。你知道我们的感官和车的反应触发器之间的直接联系：声音、速度、视觉吗？想象一下那种感觉，还有更多的感受。我对未来很乐观，汽车产业和技术公司将会保持汽车的魅力——毕竟这是一个身临其境的体验。

但是，如果你不喜欢这个车联网的曙光时代，那么你现在就可以购买你的（不联网的）梦想的汽车了。

——《连线》杂志，洛夫斯基

第 6 单元

课文 A　量子通信“跳出”实验室

——随着“现实世界”试验成功地传送了量子密钥和数据，中国开始致力于研究超级安全的网络。

量子物理学使得网络安全与远距离安全地传送数据的梦想更接近了一步——它受到了两方面进展的推动。

本周，中国将开始安装世界上最长的量子通信网络，其中包括北京和上海之间约 2000km 的链接。本周，东芝公司、英国电信公司、爱德华公司以及位于特丁顿的英国国家物理实验室共同宣布了一项研究成果，报道了来自网络现场试验的“令人振奋”的结果，表明量子通信在现有的光纤基础设施上是可行的。

传统的数据加密系统依赖于“密钥”的交换（二进制的 0 和 1）来加密和解密信息。但如果在数据传输过程中黑客对该密钥进行“窃听”，那么通信通道的安全性就会遭到破坏。量子通信使用一种叫作量子密钥分配（QKD）[1]的技术，格雷戈勒 · 里波迪说，此技术是利用光子的亚原子属性来“解决当前系统的最薄弱环节”，他是 ID Quantique 公司的联合创始人兼首席执行官，该公司是瑞士日内瓦城的一家量子密码公司。

该方法允许用户发送被置于表征密钥的特定量子态的光子脉冲。如果有人试图截听密钥，那么窃听的行为会在本质上改变其量子态——提醒用户存在安全漏洞。无论是 1 亿美元的中国首创，还是用最新研究进行测试的系统都在使用 QKD。

位于合肥的中国科技大学的量子物理学家、中国项目的带头人潘建伟说，中国的网络

“不仅为政府和金融数据提供了最高水平的保护，还为量子理论和新技术提供了一个试验平台”。

潘希望利用网络和他的团队计划在明年发射该量子卫星来验证这种想法。总之，潘说，该技术能够进行基础量子理论的进一步大规模测试（约2000km），如量子非定域性[2]。在这种情况下，改变一个粒子的量子态会影响另一个粒子的状态，即使这两个粒子相距很远。

远距离地发送单光子是QKD中最大的问题之一，因为它们往往被光纤吸收，这就使得接收终端难以检测到密钥。

海克沃·洛是加拿大多伦多大学的量子物理学家，他说，这是“传统探测器的一大挑战”。但近年来的技术突破显著降低了探测器的噪声水平，同时将光子探测效率从只有15%提高到50%。

探测器“计数”光子脉冲的速率也取得了巨大的进步——这在决定量子密钥的传送速率上是至关重要的，也因此决定了该网络的速率。洛说，计数率已经提高了1000倍，达到约2GHz。

该突破延长了量子信号的发送距离。唐·海福特说，试验使用“暗光纤”——电信公司铺设的但闲置的光纤——已经发出达100km的量子信号，唐·海福特是巴特尔一家技术开发公司的研究员，该公司总部位于美国俄亥俄州哥伦布市。

要达到更远的发送距离，量子信号必须在“节点”中继——例如北京和上海之间的量子网络将需要32个节点。不使用节点的更长距离的传输光子将需要一颗卫星。

中国并不是唯一致力于量子通信的国家。海福德带领的团队和ID Quantique公司已经开始在巴特尔的总部和华盛顿特区办公室之间安装650km长的链接了。海福特说，此次合作还规划了一个连接美国各大城市的网络，此链接可能突破10 000km，虽然此规划尚未获得经费。

中国和美国的网络都将使用暗光纤来传送量子密钥。剑桥市的东芝欧洲研究中心的量子物理学家安德鲁·希尔兹说：“但是，这些暗光纤并不是一直可用的，它们非常昂贵。”避免这个问题的一种方法是将光子流驮载到发送传统电信数据的“亮”光纤上。然而，这些传统的数据流通常比量子流强100万倍左右，所以量子数据往往会被淹没。

在本周公布的结果中，希尔兹和他的同事们成功地沿着英国电信公司两个相距26km驻地间的正在使用的亮光纤实现了QKD的稳定安全传输。量子密钥以高速率在几周内沿着同一光纤上传输强劲传统数据的四个通道进行传送。这项研究基于希尔兹和他的团队以前的工作成果，他们曾开发出一种技术来探测在一个90km长的光纤中与噪声数据一起传送的量子信号，但这是在受控的实验室条件下进行的。

希尔兹说：“在‘现实世界’实现QKD比在实验室的受控环境中实现QKD更具挑战性，这是由于环境的波动以及光纤中的更大损耗。”

在最新的研究中，量子密钥与传统数据以40Gbit/s的速度一同传送。希尔兹补充道：“据我所知，这是迄今为止QKD多路复用传输数据的最高带宽。”

他认为，沿着40个传统数据通道发送QKD信号是有可能的。光纤通常携带40~160个电信信道，这意味着量子通信可以通过现有的基础设施来进行。

海克沃·洛在一个现场实验中说：“这是首次展示在同一光纤中经典信号和量子信号的多路复用，我觉得这是一项令人印象深刻的工作。”他说，对暗光纤的需求减少是显示QKD

有潜力应用于“现实生活”的重要一步。

由 *Jane Qiu* 撰写，2014 年 4 月 23 日，来源于《自然》

课文 B　中国最新的飞跃非常了不起——这就是量子

——北京发射的世界上第一颗量子通信人造卫星进入轨道

北京——周二早上从戈壁滩发射升天的火箭将有望推动中国站在一个最具挑战性的科学领域前沿。

此举还使北京在网络间谍繁多的年代比它的国际竞争对手提前获得了一个令人高度觊觎的财富：防黑客通信。

官方媒体报道：中国于周二凌晨 1：40 在内蒙古自治区的一个发射中心发射的世界上首个量子通信卫星进入轨道。在五年的制作过程中，整个科学界和安全界都时刻对这个项目保持密切关注。

这个量子项目是中国过去 20 年数亿美元战略的最新部分，旨在在自然科学[1]研究方面追上甚至超过西方国家。

“制造量子卫星是一场巨大的竞争，很有可能中国会是这场比赛的冠军。”瑞士日内瓦大学的教授及量子物理学家尼古拉斯·吉辛表示，“这再一次表明中国有能力开展巨大且雄心壮志的项目，并有能力实现它们。”

研究人员表示，美国、欧洲、日本及其他国家的科学家们都争先恐后地开发亚原子粒子的奇特且潜能巨大的特性，但是很少能像中国的科学家们一样得到国家的支持。国家三月份公布，量子科技在中国的五年经济发展计划中是一项尖端的战略重点。

北京还没有公开在量子研究或建造这个 1400lb 重的卫星时投资了多少钱。但是用于包括量子力学在内的基础研究的资助，2015 年为 101 亿美元，比 2005 年增长了 19 亿美元。

据一些科学部、国防部、情报部及其他部门的官员在 7 月份提出的国会报告显示，美国联邦政府每年用于量子研究的投资约为 2 亿美元。报告说，发展量子科学将会“加强美国国家安全”，但又说，经费的波动已经推迟了其进程。

与此同时，北京努力吸引那些出生于中国、在国外接受教育的量子物理方面的专家回国，包括正在主导这个项目的物理学家潘建伟。

“我们从世界各地的实验室中学习了所有的好技术，我们吸收它们并将它们带回国。”潘先生周一在中国国家电视台公开的采访中说道。

有了国家的支持，潘建伟有能力超越他之前的博士生导师——维也纳大学的物理学家安东·蔡林格，从 2001 年起他就试图说服欧洲太空总署发射一个类似的卫星。

“这是一个艰难的过程，需要花费很多时间，”蔡林格先生说。他正在致力于研究他以前学生的卫星。

指导项目的潘建伟和中国科学院都未回应置评请求。欧洲太空总署也未回应。

为美国科学研究提供联邦资金的国家科学基金会称：量子科学是六“大观点”之一。物理专业小组的组长丹尼斯·考德威尔认为它是解决关键社会挑战的长期研究。她说国家科学基金会最近还设立了一项新的研究奖项，这将会直接影响到量子信息的研究。

一个在美国华盛顿特区专门从事研究中国和网络安全的研究员约翰·卡斯特罗说：中国在这个领域的投资从某种程度上说是担心美国的网络能力所致——针对 2013 年美国深入渗

透中国网络的披露。他还指出，美国的机构正在研究如何建立一个强大的量子计算机，这种计算机理论上能够破解全世界用于安全通信的基于数学的密码。

“中国政府意识到它们对电子间谍越来越敏感了。”卡斯特罗说。

然而他说，量子通信本质上就是自卫的，并且不会从美国所说的中国国家资助的黑客项目中获益。

量子加密技术是安全的，它可以对抗任何一种计算能力，因为编码到量子粒子中的信息一旦被测量，将会被毁坏。位于日内瓦的量子密码公司 ID Quantique 的联合创始人乔治·里波迪将它比作发送一条写在肥皂泡上的消息。他说：“如果有人想在它传输的过程中通过触摸将它拦截，那么这个肥皂泡就会破碎。”

量子物理学家最近提出使用光量子在地球上进行安全的短程通信。如果成功的话，卫星会极大地扩展不可破解的通信范围。为了检验量子通信是否可以在全球范围内进行，潘建伟告诉官方媒体他和他的团队将尝试通过太空从北京发送一组量子密码到维也纳。詹姆斯顿基金会[2]中研究中国情报工作的皮特说：“这些技术必然存在一些问题，人们使用这些技术时，事情往往变得很糟糕，除非他们经过大量的培训。我认为中国不仅有责任为我们自己做些什么——许多其他国家都已经去过月球，已经完成了载人航天——而且有责任探索一些未知的事物。”

“如果测试成功的话，它的影响将是巨大的，”之前在维也纳接受培训的量子物理学家，在南京大学从事卫星项目早期阶段工作的马晓松说。

新加坡量子技术中心的主要研究人亚历山大·玲说，量子密码并不是万无一失的。黑客们还是有可能通过向量子接收器发射强激光来欺骗一个不谨慎的接受者。

美国安全专家也在质疑量子通信的错综复杂之处是否可以在干扰的环境中简化使用。

维也纳大学的蔡林格说：不论有什么挑战，量子卫星都将中国和量子力学领域放置在了重要的科技突破前沿。“从长远来看，量子通信有很好的机会来取代我们现在的通信技术。”他说，“我看不到它不发生的理由。”

在一月份的《自然》杂志的采访中，潘建伟说，量子卫星显示中国的科学家已经停止追寻其他国家的步伐。为了阐释这一观点，中国官方媒体周一表示，该量子卫星以公元前5世纪反对侵略战争的哲学家“墨子”命名。

《华尔街日报》的乔希·覃报道，薇薇安·庞撰写文本

第7单元

课文A　存储程序控制

存储程序控制（SPC）是一种用于电话交换的通信技术，由存储于开关系统内存中的计算机程序控制。SPC 是 20 世纪 50 年代贝尔系统（Bell System）[1]开发的电子交换系统（ESS）的使能技术。

史端乔式交换、面板、旋转、横纵开关等早期的交换技术都是完全由模控电子的机电开关元件构成的，并没有采用计算机软件控制。1954 年，美国贝尔实验室[2]的科学家厄纳·施耐德·胡佛发明了 SPC 技术，她认为可以利用计算机软件控制电话连接。

20 世纪 60 年代，SPC 被引进到电子交换系统。贝尔系统建造了机电-电子过渡阶段的交

换系统101ESS PBX，为此前仍然在使用机电化的中心交换局的商业客户提供扩展服务。当时在公共电话交换网中应用的是爱立信提供的大型系统——西部电气1ESS交换机和AXE电话交换机。SPC实现了复杂呼叫的特性。随着SPC交换机的进化，其可靠性和通用性也在增强。通过引入时分复用（TDM）技术，减少了分系统的规模，显著提高了电话网络的容量。在20世纪80年代，SPC统治了电信产业。

1. 简介

SPC的主要特点是用一个或者多个数字处理单元（存储程序计算机）执行存储在交换机系统内存中的一系列计算机指令（程序），相关电路中的电话连接由此建立、维持和终止。

存储程序控制的直接结果是交换功能的自动化，以及给用户带来多种新的通话特性。

通过引入容错设计实现了电话交换系统持续、不间断的运行。在SPC技术的早期试验中，控制子系统的电子和计算机部分均获得成功，电话交换机因此发展成为全电子系统，其交换网络也是电子的。1960年，美国贝尔实验室在伊利诺伊州莫里斯部署了SPC测试系统。该SPC测试系统利用18位字长的飞点存储器存储半永久性的程序和参数，利用内存栅障存储器随机访问运行中的内存。1965年5月，世界上第一台用于永久性生产使用的电子交换系统1号电子交换机（No. 1 ESS）在新泽西州萨克萨纳的美国电话电报公司（AT&T）服役。到1974年为止，AT&T公司已经安装了475台No. 1 ESS系统。在20世纪80年代，电信产业的机电交换被SPC替代，机电交换这一术语只剩历史价值。如今，SPC是各种电子交换机的标准特性。

利用半导体交叉接点开关取代机电交换矩阵并不是一蹴而就的，尤其是在大型交换机上。所以，许多空分交换系统在使用SPC交换系统的同时仍使用机电交换网络。但是，专用自动交换分机（PABX）和小型交换机都是使用电子交换机。

2. 类型

SPC可以通过集中式控制和分布式控制两种方式来实现。20世纪60年代和70年代开发的早期电子交换系统（ESS）几乎一成不变地使用集中式控制系统。到了现在，尽管仍有许多交换机在设计时继续采用集中式SPC，但是随着低功耗微型处理器和可编程逻辑阵列（PLA）、可编程逻辑控制器（PLC）等VLSI芯片的出现，分布式SPC在21世纪初期就已发展成了主流。

（1）集中式控制

在集中式控制系统中，所有的控制设备都由一个中央处理单元所取代。中央处理单元必须根据系统负载能力，在1秒钟的时间内处理10~100个电话。现在集中式控制系统普遍采用多处理器结构，其运行模式多种多样，譬如主备模式、同步双工模式或者负载共享模式等。

1）主备模式。对于双处理器结构的系统而言，主备模式是最简单的。通常，两个处理器中会有一个处于待机模式（即备用处理器）。只有当主处理器发生故障时，备用处理器才会上线。对于这种工作模式，最重要的一点就是当备用处理器接管控制系统时能重建交换系统的状态；这意味着备用处理器能判断出交换系统正在使用的是哪些用户线或者用户网。

在小型交换机中，备用处理器一旦被激活就可以通过及时扫描状态信号对交换系统状态进行重建。在这种情形下，只有故障出现时建立的电话连接会受到影响。而在大型交换机中，备用处理器根本无法在有效的时间范围内对全部的状态信号进行扫描。因此，主处理器周期性地将系统状态复制进辅助存储器。转换发生时，辅助存储器中最新存储的状态数据就会被加载。在这种情形下，只有那些在处理器发生故障之前且最后更新状态的电话会受到影响。共享的二级存储不需要复制，简单的单元级别冗余即可满足要求。1ESS 就是突出的例子。

2）同步双工模式。在同步双工模式中，因为硬件耦合，两个处理器能持续执行相同的指令集、持续比较指令执行结果。一旦出现不匹配的情况，在几微秒的时间内，就要识别出故障处理器并将其下线。当系统正常运行时，两个处理器的内存中一直存储着相同的信息，同步接收着交换机输出的信息。两个处理器中有一个处理器实际控制着交换网络，另一个处理器与之信息同步但是不会参与交换网络控制。如果比对器检测到了错误，两个处理器就会解耦，检测程式启动并独立运行以发现故障处理器。这个过程并不会影响电话处理，此时电话处理会暂时停止。一旦故障处理器下线，另一个处理器就必须独立运行。当故障处理器检修完毕、重新上线时，主处理器的存储内容就会复制进备用处理器的内存中，两个处理器就会同步，比对器也会重启。

此外，还存在一种可能性，那就是比对器仅因为瞬时的故障报错，检测程序根本检测不到错误。这种情形存在下面三种可能性：

- 两个处理器都继续运转；
- 将主处理器下线，让备用处理器接管；
- 让主处理器继续运转，将备用处理器下线。

第一种方案是基于假设故障是瞬时的，也许再也不会出现。在第二种、第三种方案中，需经过大量测试，判定下线处理器的微小故障。

3）负载共享模式。在负载共享模式中，来电会随机地分配给或者按预先的设定分配给处理器，而这个处理器要负责该来电的所有任务直到电话完成。因此，两个处理器同时处于活跃状态，动态地共享负载和资源。两个处理器都能访问、检测和控制全部的交换系统。因为电话是由单个处理器独自处理完成的，这些临时的电话信息就会单独存储在各自处理器的内存之中。尽管程序和半永久数据能被共享，但为了冗余它们保存在不同的内存之中。

通过处理器间链接，处理器就能交换用于相互协调的信息，验证另一方的“健康状态”。如果信息交换出现故障，检测到故障的处理器就会接管所有的负载信息，包括已经由故障处理器建立的电话。然而，那些正在由故障处理器处理的电话通常会丢失。为了防止两个处理器同时访问同一个资源，需要为资源共享建立一个排斥机制。排斥机制可以引进到软件或者硬件中，也可以同时引进。

（2）分布式控制

分布式 SPC 比集中式 SSPC 更易获得且可靠。

1）纵向分解。整个交换系统被分割成好几块，每一块都分配给同一个处理器。这个处理器要处理与特定块相关的所有任务。因此，所有的控制系统都是由几个控制单元组成的，这几个控制单元又是耦合在一起的。为了冗余目的，处理器信息也许会被复制到每一个块中。

2）水平分解。在这种类型的分解中，每个处理器仅仅执行一个或者多个交换功能。

课文B 分组交换

分组交换是一种应用于数字网络通信的技术手段，是将所有要传输的数据分成大小合适的数据块——也叫作分组，再将分组通过多个同步通信会话共享的信道进行传输。分组交换提高了网络的传输效率和鲁棒性，并实现了在同一网络上运行多种应用的技术融合。

分组由分组头和数据组成。网络硬件利用分组头中所携带的信息将分组传送到目的地，分组到达目的地后，应用软件再将用户数据提取出来使用。

20世纪50年代末，兰德公司[2]受到美国国防部的资助开展科学研究，计算机科学家保罗·贝恩[1]为了研发出容错、高效的通信信息交互方法，提出了分布式自适应信息块交换的概念。这个概念与当时已建立的网络带宽预分配原则相左并形成对比，网络带宽预分配原则很大程度上随着贝尔系统的电信业务发展而巩固。但这个概念一直无法在网络实施者中获得共鸣，直到20世纪60年代末期英国计算机科学家唐纳德·戴维斯[3]在英国国家物理实验室中独立地提出了相同的理论。戴维斯被认为是“分组交换”这个现代术语的发明人，并且在接下来的10年内启发了众多分组交换网络——包括美国早期的ARPANET也吸纳了这个理论。

1. 概念

分组交换可简单定义为：

借助于携带地址信息的分组实现数据的交互和传输，分组在传送过程中独占信道，分组传送完成后信道可用于传输其他数据。

分组交换的特点是数据流的传输速率并不是恒定的，计算机网络可以根据需要利用统计复用技术、动态带宽分配技术对分组序列的传输资源进行分配。当数据传输到网络节点（譬如交换机或者路由器等）之时，节点接收分组信息、缓冲、编队、传输（存储并向前发送），节点存储转发的吞吐量和延迟时间取决于网络的链路容量和流量负荷。

与分组交换相比，另一种应用普遍的网络范式是电路交换，其专门为每一个通信会话预分配一个专用的网络带宽，每一个网络带宽所拥有的数据传输速率是恒定的，节点之间的延时也是恒定的。在蜂窝移动通信服务等计费服务应用中，电路交换的特点是以连接时间为单位进行收费，即使没有数据传送也需要缴费；而分组交换的特点则是以传输信息量（譬如字符、分组、消息等）为单位进行收费。

分组模式应用于通信时，可以配置中间转发节点（即分组交换机或者分组路由器等），也可以不配置中间转发节点。分组通常由中间网络节点采用先入先出缓冲异步发出，但也可以根据时序安排的原则进行合理编队、流量整形、差异化或确保，例如加权合理编队或漏桶效应。倘若传输介质可共享使用（譬如无线电或者10BASE5等），那么可利用多址技术对分组进行传输。

2. 历史

20世纪50年代晚期，美国空军试图建造一个在遭遇核打击之后仍能存活的系统，以消除敌人实施首次核打击的优势，这个系统就是半自动地面环境防御系统（SAGE）。同时，美国空军为该系统建造了一个广域网络。

利奥纳德·科仑洛克在其主导的早期研究中对排队理论[4]进行了研究，排队理论被证实对分组交换理论非常重要。1961年，科仑洛克在相关的数字信息交换（没有分组）领域发表了一部著作；稍后，他参与领导了世界上第一个分组交换网络 ARPANET 的建造和管理。

在 20 世纪 60 年代早中期，美国兰德公司的保罗・贝恩和英国国家物理实验室（NPL）的唐纳德・戴维斯第一次分别独立提出了将数据分成小块再进行交换的概念。

为了让美国空军的通信网络在核袭击中存活下来，贝恩在兰德公司研究出了分布式自适应信息块交换的概念。1961 年夏天，贝恩第一次将这个想法以 B-265 简报的形式呈送给美国空军；然后在 1962 年，他又以兰德报告 P-2626 的形式将之发表；1964 年，他又发表了报告 RM3420。报告 P-2626 描述了一个大规模的、分布式的、可存活的通信网络的整体架构。这份报告着眼于三个关键概念：任意两个通信点之间构建多路径的分散式网络，将用户信息分割成信息块（后来称之为分组），将这些信息块通过存储转发交换机制进行传输。

美国信息处理技术办公室的罗伯特・泰勒和 J. C. R. 利克里德主张建造广域网，他们获悉了贝恩的研究，劳伦斯・罗伯茨受到启发，将这项技术应用到了 ARPANET 的开发之中。

从 1965 年开始，英国国家物理实验室的唐纳德・戴维斯就独立地开发出了与贝恩的理论相同的信息交互方法。戴维斯称之为分组交换，比贝恩的命名更易让人理解，他还提议在英国建立一个全国性的网络。1966 年，他发表了相关演讲，之后国防部（MoD）的一位工作人员向他介绍了贝恩的工作。1967 年，美国计算机协会（ACM）召开“操作系统原理”专题研讨会，戴维斯研究团队的罗杰・斯坎特莱伯里与劳伦斯・罗伯茨会面，建议将分组交换应用于 ARPANET 之中。

在戴维斯最初的网络设计中，他所选择的一些参数与贝恩是一样的，譬如分组的大小都是 1024B。1966 年，戴维斯建议 NPL 在实验室中建造一个通信网以满足 NPL 的需求，他还证实了分组交换的可行性。1970 年，NPL 数据通信网正式投入运行。

从 1968 年起，美国劳伦斯・利福摩尔国家实验室为了实现计算机资源共享的目的配置了第一台计算机分组交换网络，即 Octopus 计算机网络，它将四台控制数据 6600（Control Data 6600）计算机与几台存储共享设备（包括 1968 年的 IBM 2321 数据单元和 1970 年的 IBM 图片存储仪）和几百台 Model 33 ASR 电传打字机终端相连以同时使用。

1973 年，文特・瑟夫和鲍勃・卡恩为通信节点利用分组交换共享计算机资源制定了一个网络互联协议，即传输控制协议（TCP）。

3. 无连接的和面向连接的模式

分组交换可以分为无连接的分组交换和面向连接的分组交换，也可以分别称之为数据报交换和虚电路交换。

无连接传输协议的例子有：以太网协议、互联网协议（IP）和用户数据报协议（UDP）。面向连接传输协议包括 X. 25、帧中继协议、多协议标签交换（MPLS）和传输控制协议（TCP）。

在无连接传输模式中，每一个数据分组都包含完整的地址信息。每一个数据分组都经由独立路由传输，有时也会因多径传输而产生交付失序。每一个数据分组都会标识接收地址、源地址和端口数，也许还会标识分组的序列号。这就是说，无须为数据分组建立通往目的地的专用链路，但也意味着分组头中应该存储更多的信息，因此，数据头很大，这些信息需要由功耗大的按内容寻址内存器查找。每一个数据分组发送时也许是经由完全不同的路由；相比面向连接的系统在建立连接时所做的工作，无连接系统可能需要为每一个数据分组做同样多的工作，但是需要较少的信息。在目的地，原始信息/数据基于分组序列号按照正确的顺序进行重新组装。尽管中间网络节点只提供一个无连接网络层服务，但是虚拟连接（也称之为虚电路或者字节流）可通过传输层协议传输给终端用户。

在数据分组传输到目的地建立通信参数之前，面向连接的传输需要在每一个有关的节点处创建连接。数据分组中并不会携带地址信息，而是携带一个连接标识符；端点之间对分组进行协商，按照顺序传输分组，并进行误差校验。地址信息仅仅在连接创建阶段传输给每一个节点，一旦查探清楚数据分组传输到目的地的路径，就会在连接传输路径的每一个网络节点的交换表上增添入口信息。信令协议允许应用指定其需求和查看链接参数。服务参数的可接受值也是可以协商的。网络节点对数据分组进行交互时需要查看交换表中的连接 id。因为分组头只需要包含连接 id，以及长度、时间戳、序列号之类的信息，所以分组头很小。对于不同的数据分组而言，分组通常是不一样的。

第 8 单元

课文 A　信 息 安 全

信息安全[1]（有时简写成 InfoSec）是指保护信息免遭未经授权的访问、使用、披露、破坏、篡改、检查、记录和销毁。这是通用术语，不论数据的形式如何（比如电子的、物理的），这一术语都可以使用。

1. 概述

（1）IT 安全

信息安全有时称为计算机安全，信息技术（IT）安全是应用于技术（大多数时候是指某种形式的计算机系统）的信息安全。值得一提的是，计算机不一定是指台式计算机。凡是带有处理器和存储器的设备都可以称为计算机。也就是说，简单如计算器这种不联网的单机设备，到智能手机、平板电脑这种联网的移动计算设备，都可以算做是计算机。因为大型企业/公司内部的数据都具备一定的价值和性质，所以大公司都会雇佣 IT 安全专家。恶意的网络攻击会入侵公司的重要私人信息系统甚至夺取公司内部系统的控制权，IT 安全专家有责任在确保公司系统安全的基础上防止公司技术遭受此种侵害。

（2）信息安全保障

信息安全保障是指一种保证信息安全的防护性行为，防止信息的机密性、完整性和可用性（CIA）被侵犯。譬如，保证信息即使遭遇重大事故也不会丢失，此类重大事故包括但不限于：自然灾害、计算机/服务器功能障碍或者物理偷窃等。因为在现代社会中，大多数信息都存储在计算机里，所以通常是由 IT 安全专家来保障信息安全。保障信息安全的一种常见方法就是为数据提供非现场备份，以防万一发生上述所提及的事故。

（3）威胁

信息安全面临的威胁来自多个方面，其中最常见的有软件攻击、知识产权剽窃、身份盗用、设备或信息被盗、信息破坏以及信息敲诈等。大多数人的电子设备都遭受过网络软件攻击，其中比较常见的形式是网络病毒攻击、蠕虫病毒攻击、钓鱼攻击、特洛伊木马攻击等。对于 IT 领域的许多企业而言，知识产权剽窃也是非常普遍的安全威胁。身份盗用是一个人尝试以他人的身份获取其个人信息或者利用他人的访问权访问重要信息。在当今社会，电子设备日趋移动化，移动电话的信息容量逐渐增大，手机被盗的风险也越来越高，这使得设备或信息被盗变得越来越普遍。信息破坏通常是指企业网站被非法破坏，使得企业在顾客那里的信誉度受损。信息敲诈是指不法分子利用勒索软件盗窃公司的信息或者资产然后要求公司进行赎买的行为。预防网络软件攻击有许多种方法，但是用户谨慎小心才是最有效的预防措施之一。

政府、军队、企业、金融机构、医院和私人企业收集了大量机密信息，涉及其雇员、客户、产品、研发和财务状况等多方面。大部分信息都是在电子计算机上进行收集、处理和存储的，还会通过网络共享给其他计算机。

如果顾客的机密信息或者公司的财务状况或者新产品生产线资料落入竞争对手或者黑客之手，会给公司和公司顾客带来巨大的、无可挽回的经济损失，并且损害公司声誉；但就公司的角度而言，保障信息安全也需要平衡成本。因此，科学家为解决此类问题设计了一个数学经济学方案，即戈登-洛布模型[2]。

对个人而言，信息安全对个人隐私影响重大。当然，在不同文化中，对隐私的看法也千差万别。

近几年，信息安全领域得到了显著壮大和发展，并为多个领域提供专业服务，譬如网络和基础设施维护、应用和数据库维护、安全测试、信息系统审计、业务连续性计划和数字取证等。

（4）应对信息安全威胁

面临信息安全威胁或者信息安全风险的应对策略可能有：

- 减少/缓解——实施安全保障措施与应对策略，消除安全漏洞、拦截安全威胁。
- 指派/转让——将安全威胁成本转嫁给另一个公司或者企业，譬如购买保险或者将服务外包。
- 接受——评估实施应对策略的成本是否超过安全威胁可能造成的成本损失。
- 无视/拒绝——既无效又不明智的应对之法。

2. 定义

在不同的文件中，信息安全的定义各有不同，总结如下（援引）：

1）“保护信息的机密性、完整性和可用性。注释：另外，其他诸如真实性、可核查性、不可否认性和可靠性等性质也包括在内。”（国际标准 ISO/IEC 27000：2009）

2）“为了保护信息的机密性、完整性和可用性，保护信息和信息系统免遭未经授权的访问、使用、披露、破坏、篡改和销毁。”（国家安全系统委员会 CNSS，2010）

3）“确保只有授权用户（机密性）在有需要时（可用性）能访问准确的、完整的信息（完整性）。”（信息系统审计与控制协会 ISACA，2008）

4）“信息安全是企业知识产权的保护过程。”（D. 派普金，2000）

5）“……信息安全是一门风险管理学科，主要研究企业应对信息安全风险的管理成本。”（E. 麦克德莫特 &D. 吉尔，2001）

6）“深入了解信息风险和信息控制平衡之法的一种保障概念。”（J. 安德森，2003）

7）“信息安全是保护信息、最大限度地减少其暴露给未授权方的风险的行为。”（H. S. 文特尔 &J. H. P. 伊洛夫，2003）

8）“信息安全是一门涉及多学科研究领域的专业学科，为了将信息保护在其应有的位置上（企业内部或者企业外围），研发所有可利用的（技术上的、组织上的、以人为导向上的、法律上的）安全机制模式并应用到实践中，以保护对信息进行创造、处理、存储、传输和破坏的信息系统免遭安全威胁。将信息和信息系统所遭受的威胁进行分类，再根据每一种威胁所属的类别进行分析、定义其相应的安全目标。该安全目标应定期进行修订，以确保其能与不断变化的环境保持充分一致。在现阶段，与信息相关的安全目标可能包括：机密性、完整性、可用性、隐私性、真实性及可信性、不可否认性、可核查性和可审查性。”（Y. 谢尔丹特塞瓦 &J. 希尔顿，2013）

3. 基本准则

（1）主要概念

信息安全的核心是 CIA 三要素，即机密性、完整性和可用性。（在文献中，信息安全经典三要素——机密性、完整性和可用性——可以交替归类为安全属性、安全特性、安全目标、信息基本要素、信息标准、信息关键特征和信息基本构造模块等。）关于是否要扩展这个经典三要素，一直存在争议。可核查性等其他准则有时也会被建议添加进来——已经明确的是不可否认性并不适合在三大核心要素之内。

经济合作与发展组织（OECD）在 1992 年提出并于 2002 年修订了《信息系统和互联网安全》九大通用准则，即意识、责任、响应、伦理、民主、风险评估、安全设计 & 实施、安全管理和再评估。基于此，2004 年，美国国家标准与技术研究院（NIST）提出了《信息技术工程安全》33 准则，并从每一条准则中派生出了指导原则和通用做法。

2002 年，唐・帕克提出，用“信息六元素”替代经典的 CIA 三要素。这六个元素分别是机密性、占有性、完整性、真实性、可用性和实用性。该标准价值几何？安全专家尚无定论。

2011 年，国际开放标准组织（Open Group）公布了信息安全管理标准 O-ISM3。该标准就安全的主要概念提出了一个可操作型定义，即与访问控制、可用性、数据质量、符合性及技术性等元素相关则可称之为“安全对象”。这个标准目前并未得到广泛认可。

（2）机密性

在信息安全中，机密性“是使信息不泄露给未授权的个体、实体、过程或不使信息为其所用的特性”。（摘录自 ISO27000）

（3）完整性

在信息安全中，数据完整性是指在数据整个生命周期内，维护和确保数据的准确性和完备性。这意味着未经授权或者未经检测，数据无法被修改。这与数据库的参照完整性不一样，尽管可以将信息完整性看作是 ACID[3] 经典模型在交易处理过程中维持数据库一致性的特殊情况。通常，信息安全系统不仅要维护数据的完整性也要维护数据的机密性。

（4）可用性

对于任何具备可用性的信息系统，在对信息有需求时信息就必须可用。这意味着，存储和处理信息的计算机系统、保护信息安全的控制系统、访问信息的通信信道都必须保持正常运行。高可用性的系统致力于在任何时候都能保持信息的可用性，即使因停电、硬件故障和系统升级而导致服务中断时信息仍然可用，即使出现阻断服务攻击——譬如目标系统接收大量输入信息从而在实质上迫使系统关闭——信息仍然可用。

(5) 不可否认性

在法律上，不可否认性针对的是一个人履行合同义务的意愿。也是指交易的一方不能对接收交易进行否认，另一方也不能对发送交易进行否认（这也可以看作是“部分完整性”）。

需要重点指出的是，密码系统等信息技术能在交易参与者的不可否认的行为中发挥重要作用，但在不可否认性的核心部分，法律概念超越了技术概念。例如，信息技术并不足以证明信息上签署的签名密钥与信息发送方是匹配的，也不能证明只有信息发送方能发送信息、其他任何人都无法在信息传输过程中更改信息。相反，这个所谓的信息发送方可以证明电子签名算法易受攻击且是有缺陷的，或者声称或者证明他的签名密钥受损了。造成这些违规的缺陷有可能在于也有可能不在于信息发送方自己的责任，这样的主张有可能会也有可能不会解除信息发送方的责任，但是这个主张将使得“签名必定证实信息的真实性和完整性并因此实现信息的不可否认性”的说法失效（因为权威性与完整性是不可否认性的前提）。

课文 B 2016 年五大主导信息安全趋势

似乎每一年，网络犯罪带来的威胁都会进化出新的更危险的形式，而网络安全机构只能拼命追赶。

随着 2015 年接近尾声，我们可以预期，网络犯罪活动的规模、严重性和复杂性都将在 2016 年继续增加，信息安全论坛（ISF）现任总经理史蒂夫·德宾如是说。ISF 是一个非营利性企业联合会，为其成员评估安全和风险管理问题。

“在我看来，2016 年可能是网络安全风险最为严峻的一年。”德宾说：“我之所以这样说，是因为我认为我们越来越多地认识到这样一个事实：网络运营有自己的特点。”

“当我们进入 2016 年，网络攻击方式将会继续变得更具有创新性且更为复杂。”德宾说：“不幸的是，企业在开发新的安全管理机制，网络犯罪也在研究新的规避技术。在推动企业网络更具弹性的过程中，企业需要将风险管理重点从纯粹地确保信息数据的机密性、完整性和可用性扩展到企业声誉和客户渠道保护等方面，而且安全机构还应该意识到网络空间活动造成的意想不到的后果。安全机构只有对未知的各种突发情况做好准备，才能拥有足够的应变能力，以抵御各种意想不到的、高冲击性的网络安全事故。”

1. 政府干预网络所带来的意外后果

德宾说，2016 年，官方干预网络空间所带来的冲突性，将会对所有依赖网络的机构造成间接伤害，甚至造成不可预见的影响和后果。他指出，改变监督和立法将有助于限制这些活动，无论其是否以攻击企业为目标。他警告称，不牵涉违法活动的企业网站也会遭受间接伤害，因为当权者监督着网络上的每个角落。

“我们已经看到欧洲法院判决《安全港协议》[1]无效，”德宾说：“同时，我们看到政府正越来越多地要求软件设置后门程序，尽管某些技术供应商会说：‘我们是幸运的，因为我们将网络上从一个终端到另一个终端的所有信息都进行了加密，我们也不知道这些信息是些什么。’在这个由一条信息物理链接连接的世界里，连网络恐怖主义行事起来都变得日趋规范，面对这一问题，我们该如何立法呢？”

德宾说，为了继续向前发展，企业需要弄清楚政府的监督要求是什么以及和同行展开合作。“立法者需要一直跟紧最新网络攻击技术的步伐，我甚至认为立法者本身需要提高他们的业务水平，”德宾说：“立法者所探讨的一直是如何应对昨天的网络攻击事件，但是网络安全谈论的是明天的事。”

2. 大数据[2]将带来大问题

现如今的企业在运营和决策过程中，正越来越多地运用到大数据分析。但是它们必须认识到，数据分析中是存在人为因素的。德宾说，对于那些没有尊重人为因素的企业而言，高估大数据输出的价值，将置企业于危险境地。德宾指出，信息数据集的完整性较差，将很可能会影响分析结果，并导致糟糕的业务决策，甚至造成商业机遇的错失、企业品牌形象的受损和利润的损失。

“当然，大数据分析是一个巨大的诱惑，而当您访问这些数据信息时，必须确保这些数据信息是准确的。”德宾说道：“于我而言，关乎数据完整性的问题是一个大问题。当然，对于如今的企业而言，数据就是血液，但我们真能知道是A型血还是O型血吗?”

“现在企业已经收集了大量的信息，”他继续说道：“最让我担心的一件事并不是犯罪分子窃取信息，而是他们以不为外人所知的方式操纵信息。”

例如，他指出，企业外包[3]代码编程已经好几年了。“我们并不能确切地知道，这些代码中是否存在能泄露信息的后门。”他说道：“事实上，后门很可能是存在的。你更需要怀疑的是：不断地假设问题，确保代码信息确实如其所说。”当然，你不仅仅需要担心代码的完整性，还需要弄清楚所有数据的来源。“对于我们自己收集的信息，我们知道出处，这很好。”他说道：“但一旦你开始将你的信息共享，你就打开了自己。你需要知道这些信息是怎么被利用的，是与谁共享的，谁增添了新信息以及这些信息是如何被操纵的。”

3. 移动应用和物联网

“智能手机和移动设备的普及使得物联网（IoT）日益成为网络恶意攻击的主要目标，”德宾如是说。随着自携设备使用人数的快速增长和可穿戴技术引入到工作场合，工作和生活中移动App的高需求将会在2016年进一步增加。为了满足这种增长的需求，软件开发者在面临高压力和微薄利润空间的情况下，会力图缩短交货时间和降低成本。他们会牺牲安全性能，在未完全通过全面测试的情况下，推出更易被犯罪分子或者黑客所攻击的劣质产品。“不要将这与手机搞混淆了，”德宾说：“移动性不只是手机。智能手机只是移动性的一个组成成分。”他指出，像他这样总是需要出差办公的人越来越多。“就其本身而论，我们没有固定的办公室，”他说道：“我上一次登录网络是在一家酒店。而今天则是在别人的办公场所。我如何确保是我‘史蒂夫’本人登录了这个特殊的系统?我可能知道这是史蒂夫的设备登录的，或者我相信是史蒂夫的设备登录的，但是我怎么知道在设备另一端的一定是史蒂夫呢?”

“企业应该为拥抱越来越复杂的物联网做好准备，而且还要弄清楚物联网对它们意味着什么”，德宾如是说。首席信息安全官（CISO）[4]应该积极主动，确保企业内部开发的App能遵守公认的系统开发生命周期方法，严格按照测试步骤进行测试，为企业面对那些不可避免的未知情况做准备。企业也应该按照现有的企业资产管理流程和政策管理员工的设备，将员工设备纳入现有的资产管理标准之中，以创新的方法提高员工的BYOD风险知识和意识。

4. 网络犯罪带来完全的威胁风暴

“2015年，网络犯罪占据人类所面临的安全威胁排行榜的榜首，到了2016年，这种情况也不会得到解决”，德宾如是说。网络犯罪和黑客活动的增加，企业为应对监管需求的上涨而导致的成本激增，企业为防范安全部门投资下降而不懈追求技术进步，这些因素结合起来，可能形成最完美的安全威胁风暴。那些采用了风险管理办法的企业能确定自己的业务最依赖的技术是什么，从而游刃有余地对其业务进行量化，实现弹性投资。

渴望赚大钱的犯罪分子、活跃分子和恐怖分子都视网络空间为一个越来越具有吸引力的狩猎场，他们通过网络袭击破坏甚至弄垮公司和政府机构网络。企业必须为这种不可预知的局面做好准备，才能灵活地抵挡住这种不可预料的、高冲击性的网络事件。

“我看到，越来越多日益成熟的网络犯罪团伙，”德宾说：“他们出乎意料地组织复杂，协调良好。我们看到，网络犯罪作为一种服务正日趋增多。这种日益增长的复杂性会给企业带来真正的挑战。在我们正在进入的这个领域，你根本不能预测到网络犯罪是如何找到你的。从企业的角度来看，你要如何防御呢？”

部分问题在于，许多企业仍然着重于保护企业网络的外部边界，但在这个年代，企业内部的人士——无论其是出于恶意目的或者只是无知而采用了不合理的安全实践措施——正使得企业网络边界的防渗透能力越来越差。

“不管是对还是错，我们都是从外部攻击的角度来看待网络犯罪的，所以我们试图在外围构建一个安全墙来防御，”德宾说。“但是企业内部也存在威胁。从企业角度来看，这让我们非常不舒服。”

事情的真相是，除非企业采取更具有前瞻性的手段，否则它们无法解决网络犯罪难题。

德宾说：“几周之前，我与一名在大公司拥有9年工作经验的CISO交谈时，他告诉我，借助大数据分析，他现在几乎能对整个企业实现完全可视化。这是在9年后。但是网络犯罪分子很久之前就拥有了这种能力。而我们的安全措施一直都是被动防御，而不是主动出击。”

“网络犯罪分子的工作方式不是这样的，他们不是基于历史，”他继续说：“他们总是试图想出一种新方法。我认为，我们在防御战中仍然表现得不够好。我们真的需要将自己的技术水平提高到与他们同一等级。我们永远不会像他们那样富有创造力。我们公司内部仍然存在这种想法：既然我们没有被攻克过，那么为什么要花这些钱呢？”

5. 技术差距将成为信息安全的一个深渊

随着网络攻击活动越来越复杂，信息安全专家正变得越来越成熟，企业急需的信息安全专家也变得越来越稀缺。网络犯罪分子和黑客分子的数量越来越庞大，技能越来越娴熟，“红客”正奋起直追，德宾如是说。CISO需要在企业内部构建可持续性的人才招募计划，培养和留住现有的技术人才，提高企业网络的防御应变能力。

“在2016年，随着网络超链接越来越强，这种情况将变得更糟，”德宾如是说。CISO在帮助企业及时获得新的技能方面需要变得更积极。

“2016年，我们将越来越多地认识到，企业在安全部门也许并没有合适的人才。”他说道：“我们知道，我们在修复防火墙等方面招揽了一些很优秀的技术人才。但是最好的技术人才也能干涉业务挑战和业务发展的网络安全。这是一个明显的弱点。企业董事会将会意识到，企业网络是他们做生意的重要方式。我们仍然没有在业务和安全实践之间建立起连接。”

有时，企业会明显发觉，安全部门内部并没有足够优秀的CISO。有时，企业也需扪心自问，企业内部的关键部门是否足够安全。

“您的企业无法避免每一次严重的网络安全事件，虽然许多企业在网络突发事件管理方面非常完善，但很少有企业已经建立了一套有组织的方法对‘出错内容’进行评估，”德宾说：“因此，这会产生不必要的成本，并让企业承担不适当的风险。企业规模各种各样，但都需要评估状况，以确保它们对未来那些新兴的网络安全事故，做好了充分的准备，且能够很好地应付。通过采用一个切实的、具有广泛基础的、协作的方式，提高网络的安全性能和应变能力，政府部门、监管机构、企业高管和信息安全专家将能更好地了解网络攻击的真实本质，以及及时恰当地进行应对。”

这篇文章出自 CIO 的托尔·奥拉夫斯路德之手。

第9单元

课文 A　多路复用和多路寻址

1. 多路复用

因为微波链路、同轴电缆链路等通信信道都造价不菲，所以通常情况下单一信道并不会只用于传输一路信号，而是会同时传输多路信号。假设信道的数据容量超过了单个用户所需的信道容量，那么多个用户就可以通过多路复用对该信道进行共享使用。多路复用是将多路信号在同一点进行局部复合再经由同一条通信信道进行传输的技术。频分多路复用和时分多路复用是最常用的两种多路复用技术。

（1）频分多路复用

在频分多路复用[1]中，多个用户通过将其信号的频率变换或者调制到不同载波上，分享同一条通信信道的可用带宽进行传输。假设载波信号的频率间隔足够宽，且调制信号的频率没有重叠，那么就能在接收端将频分多路复用的每一路调制信号都恢复出来。为了防止信号的频率重叠，也为了简化接收端的滤波器，需要设计一个防护频带（包含信道可用频谱上尚未使用的部分）将调制信号间隔开。每个用户都会分配一个指定的频段。

单个用户的信息信号可能是模拟的也可能是数字的，但是复合后的频分多路复用信号在本质上却是一个模拟波形。因此，频分多路复用信号必须经由模拟信道进行传输。远距离电话传输系统应用频分多路复用技术的历史十分悠久，譬如美国的 N 型和 L 型同轴电缆载波系统以及点对点模拟微波系统。L 波段系统采用了分级组合结构，12 个话音频带信号频分复用成一个频率范围为 60 ~ 108kHz 的群信号；5 个群信号再复用成一个频率范围为 312 ~ 552kHz 的超群信号，一个超群信号对应于 60 个话音频带信号；10 个超群信号再复用成一个主群信号。在 20 世纪 40 年代装配的 L1 型载波系统中，主群信号直接通过同轴电缆进行传输。而在微波系统中，频带信号会调制到微波载波信号上再进行点对点传输。在 20 世纪 60 年代开发的 L4 系统中，6 个主群信号复用成一个巨群信号，能携带 3600 个话音频带信号。

（2）时分多路复用

将单个信道传输信息的时间分割成不同的时间段，交织传输不同的信号，这一多路复用的过程叫作时分多路复用（TDM）[2]。只有当信道的可用数据传输速率超过所有用户的数据传输速率之和时，才能运用时分多路复用技术对这些信号进行复合传输。时分多路复用可用于数字信号和模拟信号，但在实际情况中，却几乎只用于数字信号，由此产生的复合信号也是数字信号。

在一个具有代表性的时分多路复用系统中，用户将数据提供给时分多路复用系统，扫描开关依次选择用户数据，与交织数据信号一起形成一个复合 TDM 信号。假定所有用户的数据路径与开关的扫描机构的时序一致，或者说是同步的。如果只能从每个数据源中截取一个比特，那么扫描机构将从各个数据源中选择到达的比特。然而，实际上，扫描机构通常从一路用户那里选择多个比特的数据，构成一个时隙的数据；然后扫描开关拨到下一个用户，选择另一个时隙的数据，以此类推。每一个用户都会分配一个指定的时隙。

大多数现代电信系统在远距离信号传输中运用了某种形式的时分多路复用。时分多路复用信号可通过电缆系统直接传输，也可以调制到载波信号上再通过无线电波进行传输。运用时分多路复用的系统有北美 T 型载波系统以及点对点数字微波系统。在 1962 年引进的 T1 型载波系统中，24 个话音频带信号（或者等效数字信号）时分复用成一个信号。这 24 个话音频带信号都是传输速率达到 64000bit/s、符号率达到 8000/s 的数据流，其中，每个符号对应于 8 个比特数的数据信息。时分复用过程交织传输这 24 个时隙信息，再加上一个帧同步字节，形成 193 比特帧。193 比特帧以每秒 8000 帧的速率产生，由此使得整个数据传输速率达到每秒 1.544Mbit/s。在最新型的 T 型载波系统中，通常会再次使用分级方案将 T1 信号进一步复用成高数据速率信号。

2. 多路寻址

多路寻址的定义是，多路信号在同一点进行局部复合之后再经由同一条通信信道进行传输。然而，在许多情形中，信道用户的地理分布多种多样，且其通信时间也偶发、随机，所以必须对通信信道进行高效共用。在这样的情况下，人们设计了三种方案用于信道的有效共用，分别是频分多址（FDMA）、时分多址（TDMA）、码分多址（CDMA）[3]。在电话系统中，这些技术既可以单独使用，也可以结合起来使用，并组建了最先进的移动蜂窝系统。

（1）频分多址

频分多址的目标是将频谱分割成若干个频道（频隙），不同的用户占用不同的频隙传输信号。但问题是，频谱范围是有限的，潜在的通信用户却通常要比可用的频隙数要多得多。为了高效利用通信信道，FDMA 系统必须有效地管理可用的频隙。在美国广泛使用的高级移动电话蜂窝系统（AMPS）中，不同的电话用户通过 FDMA 技术使用不同的频隙。一旦一个电话完成，蜂窝基站的网络管理计算机就会将释放出来的频隙重新分配给新的电话呼叫者。AMPS 系统的一个主要用途是，只要有可能就重新安排频隙，尽可能多地容纳电话呼叫者。在一个局域蜂窝系统中，电话中断，相关频隙就会被重新分配。另外，同一个频隙在不同的蜂窝系统中可以分配给不同的电话呼叫者同步使用。但是，这些不同的蜂窝系统必须在地理位置上足够分离，一个蜂窝系统的无线电信号才能在抵达另一个蜂窝系统的辐射范围时充分衰减，不影响同一频隙的同时使用。

（2）时分多址

时分多址的用途是将信道上的时间分割成时间段（时隙），不同的用户占用不同的时隙传输信号。但问题是用户的使用请求是随机的，因此时隙的请求量偶尔会多于可用时隙的数量。在这种情况下，系统必须先将信息缓冲或者存储在存储器中，直至时隙出现空余才能对数据进行传输。而信息缓冲会给系统带来延时。在 IS54 蜂窝系统中，三个数字信号利用 TDMA 技术进行交织，然后再通过信道带宽为 30kHz 的频隙进行传输——该信道在 AMPS[4]系统中只能容纳一个模拟信号进行传输。IS54 系统对数字信号进行缓冲再进行时分交织，这会导致额外的延时，但是该延时十分短暂，人们在打电话时通常很难留意到。IS54 系统同时使用了时分多址和频分多址技术。

(3) 码分多址

在码分多址系统中，多路信号是在同一时间经由同一频带发送的。在接收端，用户通过其特有的签名波形对信号进行筛选或者驳回。这个签名波形是由指定的扩频代码构建的。IS95 移动蜂窝系统应用了码分多址技术。在 IS95 系统中，首先将发送给蜂窝基站的模拟语音信号进行量化，然后将之组织成数字帧结构。在一个帧结构中，一个持续时间为 20ms 的帧包含了 192B。在这 192B 中，有 172B 是语音信号本身，有 12B 是循环冗余校验码，可用于误差检测，有 8B 是编码器“尾巴”，可用于解码器进行正确解码。所有这些字节构成编码数据流。将编码数据流进行数据交织，然后每 6B 一组进行分组。一组 6B 表明可传输 64 种波形，每一种传输波形的极性交替模式都是特定的，还会占据一部分的无线电频谱。然而，在编码数据流传输之前，会将之与以 1.2288MHz 的速率进行极性交替的码序进行复用，将信号占据的带宽扩频，使之（在发射器处滤波之后）能占据大约 1.23MHz 的无线电频谱。多路用户同时使用这一 1.23MHz 的带宽，蜂窝基站根据指定的码序，就能将目标用户筛选出来。

码分多址有时会称作扩频多址（SSMA），因为利用码序对信号进行复用的过程会使传输信号的功率扩展到一个更大的宽带上。频分多址必须进行频谱管理，而码分多址则无须进行频谱分配管理。当有新的用户想要使用通信信道时，利用频分多址技术则需要对数据进行临时存储并等待频隙的释放，而利用码分多址技术则能立即给数据分配一个码序进行扩频再传输。

课文 B　正交频分多路复用

正交频分多路复用（OFDM）[1] 是一种应用于数字信号的多载频编码方法。OFDM 逐渐发展成为一种在宽带数字通信领域应用广泛的解决方案，数字电视、音频广播、DSL 上网、无线网络、电力线网络和 4G 移动通信等都运用了 OFDM 原理。

OFDM 是一种应用于多载波数字调制的频分复用（FDM）方案。OFDM 在多路并行传输的数据流或者数据通道上采用大量空间结构紧凑的正交子载波携带信号。每一条子载波信号都利用常规的调制方案（如正交振幅调制或者相移键控等）调制成了低符号率数据，使其总数据率类似于同样带宽下的常规单载波调制方案。

相比于单载波调制，OFDM 的主要优势是其有能力应对恶劣的信道条件（如长距离铜线中的高频衰减、多径传播导致的窄带干扰和频率选择性衰落等），而免去了对复杂均衡滤波器的需求。信道之所以能得到简化，在于 OFDM 可以看作是使用了许多慢调制的窄带信号而不是一个快调制的宽带信号。OFDM 信号的低符号率也使得在符号之间设计防护间隔变得划算，防护间隔不仅可以消除符号间干扰（ISI）[2]，还可以利用回声和时间扩展（在模拟电视机上分别表现为重影和模糊）实现分集增益——也即优化信噪比——的目的。这种机制也有利于单频网络（SFN）的建设——几个相邻的发射器在同一频率同步发送相同的信号，因为与传统单载波调制系统中普遍存在干扰不同，从多个远程发射机发射而来的信号可以很好地合并起来。

1. 应用举例

下面列出了所有现有的基于 OFDM 的标准和产品的摘要。

有线

- 经由 POTS 铜线传输的 ADSL、VDSL[3] 宽带接入；
- 有线数字电视标准 DVB-C 的增强版 DVB-C2；
- 电力线通信（PLC）；
- 为现有的家庭有线（电力线、电话线和同轴电线）提供高速局域网络的标准 ITU-T G. hn；
- TrailBlazer 电话线调制解调器；
- 应用于家庭网络的同轴电缆多媒体联盟标准（MoCA）；
- DOCSIS[4] 3. 1 宽带配送。

无线

- 无线局域网（WLAN）接口标准 IEEE 802. 11a，g，n，ac 和 HIPERLAN/2；
- 数字广播系统 DAB/EUREKA 147、DAB +、DRM、高清晰度无线电、T-DMB 和 ISDB-TSB；
- 地面数字电视系统 DVB-T 和 ISDB-T；
- 地面移动电视系统 DVB-H、T-DMB、ISDB-T 和 MediaFLO 前向链路；
- 基于超宽带（UWB）的无线个人局域网（PAN）标准 IEEE 802. 15. 3a，由无线媒体联盟提出并实现。

基于 OFDM 的多址技术 OFDMA 也应用于多种准 4G 和 4G 蜂窝网络系统和移动带宽标准之中：

- 移动模式的无线城域网（MAN）/宽带无线接入（BWA）（或者移动无线城域网）的标准 IEEE 802. 16e；
- 移动宽带无线接入（MBWA）标准 IEEE 802. 20；
- 3GPP 长期演进（LTE）四代移动宽带标准的下行链路。其无线接口曾被命名为高速 OFDM 分组接入（HSOPA），现在被命名为演进的 UMTS 地面无线电接入（E-UTRA）。

2. 主要特征

优点总结

- 与扩频等其他双边带调制方案相比，OFDM 具有高频谱效率；
- 无须复杂的时域均衡器就能应对恶劣的信道环境；
- 对窄带信道中的同信道干扰具有良好的鲁棒性；
- 对符号间干扰（ISI）和多径衰落具有良好的鲁棒性；
- 能通过快速 FFT 高效实现；
- 对时间同步误差具有较低的敏感性；
- 无须配置调谐的子信道滤波接收机（与常规 FDM 不同）；
- 有利于单频网络（SFN）（即发射机可以宏分集）。

缺点总结

- 对多普勒频移比较敏感；
- 对频率同步问题比较敏感；
- 峰值平均功率比（PAPR）比较高，需利用线性发射机电路，其功率效率较低；
- 循环前缀/保护间隔导致功率损失。

正交性

从概念上讲，OFDM 是一种很特别的 FDM，附加的约束就是：所有载波信号是相互正交的。

在 OFDM 中，子载波频率是这样挑选的，即子载波信号频率正交，这意味着子信道之间的串道会消除，载波间的防护频带也可以舍弃。这与常规的 FDM 非常不同，这极大地简化了发射机和接收机的设计，而且无须为每个子信道设计单独的滤波器。

正交子载波间距是 $\Delta f = \frac{k}{T_U}$（Hz），其中，T_U 是有用符号的持续时间（接收端的窗口大小），k 是一个正整数，通常为 1。因此，N 个子载波信号，其总通带带宽为 $B \approx N \cdot \Delta f$（Hz）。

正交性还能提高频谱的效率，使等效基带信号的总符号速率接近奈奎斯特采样率（即几乎是双边带物理通带信号的奈奎斯特采样率的一半），几乎使全部的可用频带都能得到利用。OFDM 通常有一个近似于白频谱的特性，从而使它与其他共享通道的用户相比，具有良好的电磁干扰特性。

一个简单的例子：一个持续时间的有用符号 $T_U = 1\text{ms}$，就需要间距为 $\Delta f = \frac{1}{1\text{ms}} = 1\text{kHz}$（或者其整数倍）的正交子载波信号。$N = 1,000$ 个子载波信号将导致总通带的带宽为 $N \cdot \Delta f = 1\ \text{MHz}$。理论上，根据奈奎斯特采样定律，该符号时间所需带宽应该为 $BW = \frac{1}{2T_U} = 0.5\text{MHz}$（即方案所需为可达带宽的一半）。如果再施以防护间隔，所需奈奎斯特带宽将更小。FFT 将导致每符号采样 $N = 1,000$ 个样本。如果不再施以防护间隔，将导致一个复数值的基带信号采用 1MHz 的采样率，根据奈奎斯特采样定律，其基带带宽就需要 0.5MHz。然而，射频通带信号是由一个基带信号和一个载波信号（即双边带正交调幅）相乘而得，其通带带宽为 1 MHz。而采用单边带（SSB）或者残留边带（VSB）的调制方案，则就相同的符号速率，所获得的带宽几乎为一半（即，就相同的字母符号长度，频谱效率高达两倍）。然而，后者对多径干扰更加敏感。

OFDM 要求发射器和接收器之间能实现精准的频率同步；倘若频率不同步，则子载波就不再正交，将会导致载波间干扰（ICI）（即子载波间的串扰）。通常，发射器和接收器的振荡器不匹配，或者运动引起的多普勒频移，都会导致频率偏移。多普勒频移可以由接收器进行补偿矫正，但是倘若结合多径技术将使情况更加糟糕，因为各种频谱偏移都会出现反射，这就更难矫正了。通常，这种效应会随着运动速度的提高而恶化，这也是高速机动中限制使用 OFDM 技术的主要原因。为了缓解类似情形下的 ICI[5]，一种方法是对每一个子载波信号都进行整形以将导致非正交子载波信号频率重叠的 ICI 最小化。例如，低复杂度的方案 WCP-OFDM（加权循环前缀正交频分复用）为了执行一个潜在的非矩形脉冲整形和利用单抽头子载波均衡器实现接近完美的信号重建而在发射机输出处使用了短滤波器。其他 ICI 抑制技术通常会显著提高接收器的复杂性。

第 10 单元

课文 A　调制与解调

在很多电信系统中，都必须将信息承载信号表达成传输媒介能精确传输的波形。而这个适合传输的波形就是通过调制来完成的。调制就是将载波的波形特征改变成与信息信号或者调制波形一致，然后利用信道传输调制信号，再通过解调过程将原本的信息承载信号恢复出来。调制应用于信息信号是有很多原因的，其中几条概括如下：

1）许多传输信道都具有通带有限的特征——即，在没有发生严重衰减（信号幅度降低）的情况下，信道只能传输一定频率范围内的信号。因此，必须应用调制技术将信息信号的“频率变换”到信道能传输的频率范围上。交流电耦合同轴电缆和光纤电缆都具备通带有限的特征，前者只能传输频率范围介于60kHz到几百MHz的信号，后者只能在没有大幅度衰减的情况下传输指定波长范围内的光信号。在这些例子中，就需要运用频率翻译手段将信息信号的频率变得“适合”信道传输。

2）在许多实际应用中，一个通信信道并不只是传输单路信号，而是由多路用户共享使用。为了防止相互干扰，每个用户都会分配一个特定频率的载波，用于调制其信息信号。分配和调制完成之后，紧接着在中心点对这些信号进行复合，复合的结果就是频分复用信号。倘若频率上没有中心复合点，那么通信信道本身就会表现为一个分布式复合器。广播无线电频段（频率范围为540kHz到600MHz）就是后者这种情况，能同时传输调幅广播信号、调频广播信号和电视信号，只要每个信号都分配了不同的频段就不会发生相互干扰。

3）即使通信信道能直接传输信息承载信号，也有一些应用上的理由导致直接传输是不可取的。举一个简单的例子：通过无线电波传输3kHz的语音频段信号。在空旷的空间中，3kHz信号的波长是100km。而一个有效的无线电天线的长度通常是信号波长的一半，一个3kHz无线电波就需要一个长达50km的天线。在这种情况下，将语音信号的频率搬移到一个较高的频率上，就只需要一个长度小很多的天线。

1. 模拟调制[1]

正如模数转换中强调的，语音信号以及音频信号、视频信号本质上是模拟信号。在大多数现代通信系统中，这些信号在传输之前都需要进行数字化，但是在某些系统中，无须将模拟信号转换成数字形式就能直接传输。常用的模拟信号调制方法有两种：一种称为调幅，使载波的幅度按照需要传输的信息信号的变化规律而变化，而载波频率保持不变。另一种称为调频，使载波的频率按照需要传输信息信号的变化规律而变化，而载波幅度保持不变。

2. 数字调制[2]

为了在通信信道上传输计算机数据等数字信号，需要将模拟载波信号调制成反应数字信号双极性特征的数字基带信号。载波参量中能调制的是幅度、频率和相位。

3. 幅移键控[3]

如果载波参量中唯一被信息信号控制变化的是载波幅度，那么这种调制方法就叫作幅移键控（ASK）。ASK能看作是模拟信号调幅的数字版本。在最简化的ASK版本中，射频信号仅仅在二进制信号为1的状态下接通传输，在二进制信号为0的状态下则断开传输。在其他ASK的改进版本中，调制信号中的0和1分别表示成两个预选振幅之间的切换。

4. 频移键控[4]

如果信息信号选中频率来控制载波变化，那么这种调制方法就称为频移键控（FSK）。在最简化的FSK信号版本中，数字信号传输时表现为两种频率形式：一个频率用来传输数字信号1；一个频率用来传输数字信号0。贝尔公司在1962年引入的103语音频带调制解调器中运用了这种调制方法，使得公共交换电话网络上的信号最高传输速率达到了300bit/s。在贝尔103调制解调器中，在发送、接收方向上，二进制数据的传输频率分为1080 +/ -100Hz和1750 +/ -100Hz。

5. 相移键控[5]

如果信息信号选择控制的参量是相位，那么这种方法就叫作相移键控（PSK）。在最简化的PSK版本中，一个相位固定的射频载波信号用来传输数字信号0，一个180°相移（即相反的极性）的射频信号用来传输数字信号1。贝尔公司在1980年左右引入的212调制解调器中运用了PSK技术，使得公共交换电话网络上的信号最高传输速率达到了1200bit/s。

6. 先进方法

除了上面所介绍的几种基本的数字调制形式外，还存在将多个调制信号叠加的改进方法。正交调幅（QAM）就是其中一种调制形式。QAM信号实际上是在两个相位正交（即相位相差90°）的载波信号上进行调幅，合并信号的每个位移都代表4个或者4个以上的比特信息。大多数运用QAM技术的语音频带调制解调器传输信号的速率超过2400bit/s，实际应用包括美国和日本的数字移动蜂窝通信系统。

格状编码调制（TCM）是一种将卷积码和QAM技术结合起来的调制方法。现如今，大多数传输速率位于9600bit/s及以上的语音频段调制解调器都是以TCM技术为基础的，譬如V.32和V.34调制解调器等。

课文B　脉冲编码调制

脉冲编码调制（PCM）是一种对模拟信号进行数字化采样的技术。在计算机、光碟机和数字电话等应用的数字音频中，其标准格式就是PCM信号。对模拟信号的振幅在均匀间隔上进行有规律的采样，再对采样结果进行量化，量化成数字电平范围内最接近的数值，形成PCM数据流。线性脉冲编码调制（LPCM）是PCM技术的特殊形式，其量化电平是线性均匀的。在这一点上，LPCM技术与PCM编码技术是相反的，PCM编码技术的量化电平随着振幅的变化而变化（正如A律压缩算法和μ律压缩算法）。尽管PCM是一个更为广义的术语，但是PCM还是经常被用来描述LPCM编码数据。

一个PCM数据流有两大基本属性，来确定其对原始模拟信号的保真度，这两个属性分别是：采样率，即1s内的采样次数；位深度，即每个采样点用几个比特数值来表示。

1. 实现

尽管有脉冲密度调制[1]（也应用于超级音频CD）等编码技术可用，但无压缩音频通常还是应用PCM编码技术进行编码。

1976年，美国基于中等规模集成电路技术，利用4ESS交换机将时分切换引入到了电话系统之中。

1982年，光碟（俗称音频CD）红皮书标准中LPCM用于音频数据的无损编码。

AES3接口标准（于1985年制定，S/PDIF标准就是基于此）是一种运用LPCM技术的特殊架构。

在个人计算机中，PCM数据和LPCM数据常常是指在WAV（定义于1991年）和AIFF音频容器格式（定义于1988年）中存储的数据。但是，其他诸如AU、原始音频格式（header-less文件）和各种不同的多媒体容器格式中也可能存储LPCM数据。

DVD标准（从1995年起）和蓝光光碟标准[2]（从2006年起）在下定义时纳入了

LPCM 技术。各种数字视频、数字音频存储格式（例如，从 1995 年起的 DV 格式，从 2006 年起的 AVCHD 格式）也是如此。

LPCM 技术也应用于 HDMI（定义于 2002 年），即用于传输无压缩数字信号的单缆数字音频/视频连接接口。

RF64 容器格式（定义于 2007 年）运用了 LPCM 技术，但也存储非 PCM 格式的比特流数据；RF64 文件中突发的各种各样的压缩数据格式（Dolby E、Dolby AC3、DTS 和MPEG-1/MPEG-2 音频）能“伪装”成线性 PCM 数据。

2. 调制

在图 10-1 中，正弦波（曲线所示）经过采样量化成为 PCM 信号。正弦波的采样间隔是均匀的，如图 10-1 中的竖线所示。对于每一个采样点而言，是利用算法计算来选择量化值（y 轴）的。点显示了输入信号完全离散化的结果，很容易就能将其编码成数字信号，用于存储或者操作。对于右侧的正弦波，我们验证采样时间点的量化数值分别为：8，9，11，13，14，15，15，15，14 等。将这些量化值编码成二进制数值，就能产生 4bit 的编码结果：1000（$2^3\times1+2^2\times0+2^1\times0+2^0\times0=8+0+0+0=8$），1000，1001，1011，1101，1110，1111，1111，1111，1110 等。可利用数字信号处理器对这些数字信号进行进一步的处理或分析。多个 PCM 数据流可以复用成一个更大的集群数据流，再利用单一物理链路进行传输。涉及的技术就有在现代公共电话系统中运用广泛的时分复用（TDM）。

通常，用一个简单的集成电路就能实现 PCM 处理，该集成电路一般被称为模数转换器（ADC）。

3. 解调

为了将原始信号从采样信号中恢复出来，解调器的操作步骤与调制过程是完全相反的。在每一个采样间隔之后，解调器都会读取下一个采样值，将输出信号位移到新值处。因为频繁转场，造成信号混叠，产生大量高频能量。为了将这些不可取的频率移除，保留原始信号，解调器必须将信号通过一个模拟滤波器，抑制设计频率范围之外（大于奈奎斯特[3]采样率）的能量。采样定理显示，如果 PCM 设备能为输入信号提供两倍于其频率的采样频率，那么在设计频段内，就不会产生频率混叠。例如，在电话系统中，可使用的语音频段范围为 300～3400Hz。因此，根据奈奎斯特-香农采样定理，为了有效重建语音信号，采样频率（8kHz）至少要是语音频率（4 kHz）的两倍。

从离散信号中精确地恢复出模拟信号的电路与根据模拟信号生成数字信号的电路很类似，这个设备就是数模转换器（DAC）。DAC 根据输入数字信号的数值生成电压或电流（取决于 DAC 类型）输出信号。输出信号一般会经过滤波、放大处理之后再使用。

4. 标准采样精度和速率

在 LPCM 中，常用的采样位深度是每个采样点用 8bit、16bit、20bit、24bit 数值来表示。单个声道做了 LPCM 编码处理。对多声道音频信号的支持取决于文件格式、LPCM 数据流的交错形式、同步方式。双声道（立体声）是应用最广的声音格式，有些设备最多能支持传输 8 声道信号（7.1 环绕）。

DVD 格式的视频通常使用 48 kHz 的采样频率，压缩盘则使用 44.1 kHz 的采样频率。一些较新的视频设备会使用 96 kHz 或者 192 kHz 的采样频率，倘若每个采样点用 16bit 数值来表示，那么双声道的最高传输速率相当于每秒 6.144Mbit，但其优势一直颇富争议。DVD 视频上的 LPCM 音频的速率极限也是 6.144 Mbit/s，即 8 声道（7.1 环绕）×48 kHz×16bit/采样点 =6，144 kbit/s。还有对每个采样点使用 32bit 数值来表示的 PCM 技术，有很多声卡支持该技术。

第 11 单元

课文 A　WiFi 如何工作

如果你最近去过机场、咖啡馆、图书馆或酒店，你就很有可能已经置身于无线网络中间。许多人在家里使用无线网络连接他们的计算机，也称为 WiFi 或 802.11 联网[1]。一些城市正在试图使用该技术给居民提供免费或低成本的互联网。在不久的将来，无线网络可能会变得很普及，以至于你可以随时随地访问互联网，而不需要使用电线。

WiFi 有很多优势。无线网络很容易建立并且价格低廉。它们也不显眼，除非你在找地方用平板电脑在线看电影，否则你都注意不到你处于一个热点中。在这篇文章中，我们将讨论使信息在空中传递的技术。我们也将讨论在你的家中创建无线网络需要什么。

1. 什么是 WiFi

无线网络使用无线电波，就像手机、电视机和收音机一样。事实上，无线网络的通信就像双向无线电通信一样，过程如下：

1）计算机的无线适配器将数据转换成无线电信号并用天线发送。

2）无线路由器接收信号并对其进行解码。路由器使用物理的、有线的以太网连接将信息发送到互联网。

该进程反向也可行，路由器接收来自互联网的信息，将其转换成无线电信号，并将其发送到计算机的无线适配器。

用于无线通信的无线电和用于无线电对讲机、手机和其他设备的无线电很相似。它们可以发送和接收无线电波，可以将 1 和 0 转换成无线电波，将无线电波转回到 1 和 0。但和其他收音机比，无线收音机有几点显著的不同：

1）它们在 2.4GHz 或 5GHz 的频率传输。这个频率对手机、对讲机和电视机使用的频率来说是相当高的。更高的频率允许信号携带更多的数据。

2）它们使用 802.11 网络标准，其中有几种：

① 802.11a 在 5GHz 发送，可移动 54Mbit/s 的数据。它还使用正交频分复用（OFDM），一种更有效的编码技术，在无线信号到达接收器之前，它们分裂成几个子信号。这大大降低了干扰。

② 802.11b 协议是最慢的、最便宜的标准。有一段时间，它的成本让它很受欢迎，但现在它不太常见了，因为更快的标准变得更便宜了。802.11b 发送在 2.4 GHz 频段的无线电频谱。它可以处理速率高达 11Mbit/s 的数据，它使用补码键控（CCK）调制来提高速度。

③ 像 802.11b 一样，802.11g 在 2.4 GHz 发送，但很快，它可以处理速率高达 54Mbit/s 的数据。802.11g 更快，因为它使用和 802.11a 一样的 OFDM 编码。

④ 802.11n 是最广泛使用的标准，和 802.11a、802.11b 和 802.11g 向后兼容。它显著地提高了速度和幅度，超过了它的前辈。例如，虽然从理论上 802.11g 速度可达到 54Mbit/s，因为网络拥塞，它真实的速度只达到 24Mbit/s。然而，据报道 802.11n 可以达到的速度

为 140Mbit/s。802.11n 最多可以传输四个数据流，每个最多 150Mbit/s，但大多数路由器只允许两个或三个流。

⑤ 2013 年年初，802.11ac 是最新标准。它尚未被广泛采用，仍以草案形式存在于电气和电子工程师协会（IEEE），但支持它的设备已经上市了。802.11ac 是和 802.11n 向后兼容的（因此其他的也是），802.11n 在 2.4 GHz 频段，802.11ac 在 5 GHz 频段。它不容易受到干扰，并且比它的前辈更快，在一个单一的数据流里最高速率可达 450Mbit/s，虽然实际速度可能较低。像 802.11n，它允许在多个空间流上传输数据，可选择地达到 8 个空间流。由于其频带，它有时被称为 5G WiFi。有时称为千兆 WiFi，因为它有可能超过每秒千兆位的多流。出于同样的原因，有时称作极高吞吐量（VHT）。

⑥ 其他 802.11 的标准可以关注无线网络的具体应用，如广域网（WAN）在车内的应用或可以让你无缝地从一个无线网络移动到另一个无线网络的技术。

3）WiFi 可以在任何三个频段上传输。或者，它们可以在不同频段迅速“跳频”。跳频有助于减少干扰，并允许多个设备同时使用相同的无线连接。

只要它们都有无线适配器，几个设备就可以使用一个路由器连接到互联网。这种连接是方便的、无形的和相当可靠的，但是如果路由器连接失败或太多的人试图在同一时间使用高带宽的应用程序，用户就会受到干扰或失去他们的连接。因此像 802.11ac 那样，新的更快的标准可以解决这些问题。

接下来，我们将看看如何从 WiFi 热点连接到互联网。

2. WiFi 热点

WiFi 热点只是一个可访问无线网络的区域。这个词最常用于公共领域的无线网络，如机场和咖啡店。有些是免费的，有些需要付费，但在任何情况下，当你忙个不停时它们都很方便。你甚至可以使用手机或外部设备连接到蜂窝网络创建自己的移动热点，而且你可以随时在家里建立 WiFi 网络。

如果你想充分利用公共 WiFi 热点或基于家庭的网络，你要做的第一件事就是确保你的计算机有正确的装置。大多数新的笔记本式计算机和许多新的台式计算机配备了内置的信号传送器，而几乎所有的移动设备都能启用 WiFi。如果你的计算机没有配备，你可以购买一个无线适配器，插入 PC 卡插槽或 USB 端口。台式计算机可以使用 USB 适配器，或者你可以买一个插入计算机机箱里的 PCI 插槽的适配器。许多适配器可以使用超过一个 802.11 标准。

一旦你安装了无线适配器和允许它运行的驱动程序，你的计算机应该能够自动发现现有的网络。这意味着，当你在 WiFi 热点中打开你的计算机，计算机会通知你，网络存在，并询问你是否要连接到它。如果你有一台旧计算机，你可能需要使用一个软件程序来检测和连接到无线网络。

能够连接到公共热点的互联网是非常方便的。家庭无线网络也很方便。你可以轻松连接多台计算机并且移动计算机的位置，网络不会中断。

3. 创建一个无线网络

如果你的家中已经有多台计算机联网，你可以创建一个无线网络与无线接入点。如果你有几台没有联网的计算机，或者你想更换你的以太网络，你就需要一个无线路由器。这个装置包含：

- 连接到电缆或 DSL 调制解调器的端口；
- 一个路由器；
- 以太网集线器[2]；
- 防火墙；
- 无线接入点。

无线路由器允许你使用无线信号或以太网电缆让你的计算机和移动设备互连、连接到打印机和互联网。大多数路由器提供全向约 100ft（30.5m）的覆盖范围，不过墙壁和门会阻断信号。如果你的家非常大，你可以购买便宜的范围扩展器或中继器，来增大你的路由器的范围。

一旦你接入路由器，它就应该以它的默认设置开始工作。大多数路由器允许你使用 Web 界面更改设置。你可以选择：

- 网络的名称，称为它的服务集标识符（SSID）——默认设置通常是制造商的名称。
- 路由器使用的频道——大多数路由器默认使用通道 6。如果你住在公寓里，你的邻居也使用 6 频道，你可能会受到干扰。切换到不同的通道可以消除问题。
- 你的路由器的安全选项——许多路由器使用一个标准的、公开可用的登录，所以这是一个好主意，设置自己的用户名和密码。

安全是家庭无线网络以及公共 WiFi 热点的重要组成部分。如果你设置路由器创建一个开放的热点，则任何有无线网卡的人都能够使用你的网络。大多数人不愿意陌生人用他们的网络。你需要采取一些安全防范措施。

重要的是要确保你的安全防范措施是最近设置的。有线等效保密（WEP）安全措施曾是广域网安全标准。其背后的想法是创建一个 WEP 无线安全平台可以让任何无线网络和传统的有线网络一样安全。但是黑客发现了 WEP 方式的弱点，如今很容易找到危及那些采用 WEP 安全模式的广域网的应用程序。WEP 由初版 WiFi 接入保护（WPA）接替，WPA 采用临时密钥集成协议（TKIP）[3]加密，是 WEP 的升级，不过也不再被认为是安全的了。

为了保持网络的私有性，你可以使用以下一种或两种方法：

- 无线保护访问版本 2（WPA2）的前身是 WEP 和 WPA，该版本是目前 WiFi 网络推荐的安全标准。
- 媒体访问控制（MAC）地址过滤与 WEP、WPA 或 WPA2 稍有不同。它不使用密码来验证用户——它使用计算机的物理硬件。

建立无线网络很简单并且不昂贵。大多数路由器的网络界面实际上是不需加以说明即可看懂的。

课文 B　NFC：开始做正事

——NFC 论坛主席田川光一

2013 年是近距离无线通信技术（NFC）在全球投入应用的重要的一年。2013 年，一些对 NFC[1]的进步至关重要的条件已经具备，包括一套全面的细则、可用的具备 NFC 功能的设备，以及一个强大的认证计划。对于那些在过去的 10 年里致力于把 NFC 带到世界各地的人来说，令人兴奋的是，“总有一天”终于变成了“今天”。在认识到这一拐点之后，NFC 论坛更加努力，在关键的垂直产业和细分市场更好地支持近距离通信方案的实施。在以下几个领域可以看到进步的迹象：

1. 具有NFC功能的各种设备正在广泛使用

根据美国的市场研究公司调查，2012年，超过1亿台NFC设备上市。预计在2013年将会超过3亿台。需要注意的是，商业化的具有NFC功能的设备种类繁多，而且十分多样化——涉及智能手机和平板电脑游戏机、笔记本式计算机、扬声器，甚至洗衣机。大多数物色NFC设备的消费者可以购买到这些产品。很多几乎不知道或者根本一点都不了解NFC的消费者将在他们的下一部智能手机或者平板电脑中获得NFC，不论他们是否能意识得到。

2. NFC的论坛规范在支持更多的功能和市场需求

2012年十月，NFC论坛批准并通过了NFC模拟技术规范。这向我们的目标迈出了重要的一步，使设备制造商更容易建立NFC论坛兼容设备的全球互操作性。因为它解决了一个NFC功能的设备的射频接口的模拟特性，也简化了测试和认证，这就是为什么我们预计它将加速引入NFC功能的设备进入市场。

一个月之后，我们发布了NFC控制器接口（NCI）的一个新技术规范，主要是在NFC控制器和设备主应用处理器之间定义NFC设备内的标准接口。NCI规范的可用性是重要的，因为它使设备制造商更容易集成来自不同芯片制造商的芯片组，并且它定义了NFC功能设备中组件之间的通用级别的功能和互操作性。随着NCI可用性的增强，制造商可以访问用于支持NFC的设备（包括手机、个人计算机、平板电脑、打印机、消费电子产品和家用电器）的标准接口。这将减轻芯片采购，并再次减少各种新型NFC功能的设备的上市时间。最近，我们宣布了新的个人健康设备通信（PHDC）规范—我们的第一个规范支持一个特定的垂直市场——保健。我们还宣布了连接切换（1.3发行版）和签名记录类型定义（2.0发行版）规范的重要更新。

3. NFC的论坛认证计划已经扩大

为了使设备制造商确保其产品符合NFC论坛规范，认证是必不可少的。我们已经增强了我们的NFC论坛认证计划，它将变得更加强大。设备制造商现在可以测试他们的产品对于数字协议、标签式经营、逻辑链路控制协议（LLCP）、简单NDEF交互协议（SNEP）[2]和模拟规范的最新版本，从而为公司带来新的NFC设备市场增添了信心和保证。

4. NFC解决方案随处可见

NFC技术的最令人满意的事情之一是其出色且多样性解决方案被推向市场。例如：

1）世界五大汽车制造商之一最近宣布，将在2015年提供一种解决方案，使具有NFC功能的设备能够与其汽车集成——用于无钥匙访问、自动个性化设置、音乐播放等。

2）2012年年底，一家排名前三的视频游戏发布商推出了一款新主机，其控制器中内置NFC功能。

3）一家领先的游戏发行商推出了具有NFC功能的视频游戏，迅速成为2012年第一大儿童游戏。

4）一个顶级的基于位置的社交网络现在支持主流智能手机平台上的NFC，让3500万用户和140万企业更轻松地分享世界各地的建议和交易。

5）美国最大的全州公共交通系统每月平均有10000个NFC支持的移动支付交易，用于铁路和公共汽车收费。

6）一家日本的在线零售商正在通过推出“购物墙”用以更快地扩张实体店，使其更易获得。“购物墙”上是独立的单元，可以让购物者能够使用 NFC 来轻碰商品，在线购买并快递到家。

7）一家瑞典与美国的合资企业推出了一个小型 NFC 功能的心电图设备，可以跟踪心房颤动患者的心律失常数据，通过 NFC 传输给医师进行监测和管理。

8）最重要的是，我的公司——索尼已经将越来越多的无线扬声器、个人计算机、智能手机和其他媒体设备中的 NFC 用于快速便捷的蓝牙和 WiFi 配对。

这些发展意味着什么？

这些只是对 NFC 做出重大战略承诺的公司的几个例子。问题的关键是，成功的创新是易传播的。当 NFC 功能的视频游戏可以成为全球顶级销售商，赢得玩具行业协会颁发的“年度最佳游戏”和“年度创新玩具”奖时，您可以确定我们还会看到其他应用 NFC 技术的游戏。当一家汽车制造商做出战略性决定，使用 NFC 作为集成和个性化使用移动设备的客户驾驶体验的手段时，其他制造商可能会效仿（实际上已经效仿了）。当认为对 NFC 的支持是社交网络必不可少的发展计划时，毫无疑问刚起步的公司也会做出同样的决策。

在所有这些例子中，公司认定 NFC 带来的附加价值是明确显著而且对企业有利的。他们相信 NFC 可以帮助他们获得竞争优势并增加收入。如果其他公司了解 NFC 的话。他们也会得出同样的结论。

现在需要什么？

早期应用 NFC 技术的公司对于证明 NFC 商业化的可行性至关重要。然而，最终 NFC 的广泛应用将取决于消费者的需求。为了增长需求，人们需要知道 NFC 是什么，以及 NFC 可以为他们做些什么。由于 NFC 支持的移动支付已经获得了重要的媒体报道，许多消费者可能已经意识到 NFC 是移动支付技术。确实，移动支付是 NFC 发展的巨大机会。根据透明度市场研究（美国市场研究咨询机构）的最新预测，到 2018 年全球手机钱包市场预计将达到 16024 亿美元，其中大部分市场的成功归功于 NFC。然而，如果我们所做的只是等待 NFC 支付业务的成熟，NFC 生态系统中的我们就没有服务于市场需求，没有为其他 NFC 服务提供基础。现在对于 NFC 设备的消费者来说，重要的是要知道 NFC 技术有许多其他例子可以对他们的生活产生直接的积极影响。

让消费者获取该消息需要一系列行动：我们需要提高消费者对 NFC 的认识以及它可以为人类做些什么。虽然有数亿个支持 NFC 的设备进入市场，但很少有 NFC 设备的功能和使用指南，或让消费者快速体验 NFC 的方式。尽管个别主流智能手机供应商把 NFC 文件共享功能作为一些电视广告的亮点令人欣慰，但是制造商还应考虑通过其他方式来提高认知度，鼓励消费者使用 NFC，包括：

1）在智能手机的包装上贴上 NFC 的标签样品，供消费者试用；

2）预装使用 NFC 的应用程序。

我们需要更多的 NFC 服务。无论是从数量还是种类上来说，NFC 的对策都在不断地增加。但对于消费者来说，商用 NFC 服务仍然相对稀缺。就像平板电脑用户触摸屏幕界面一样，当他们回到个人计算机菜单和鼠标单击时，通常会感到沮丧，习惯于 NFC 的直观触摸式界面的消费者希望能够通过快速轻触其设备来执行越来越多的操作。解决方案就是各行各业的供应商需要充分准备来满足消费者的这一需求。为了引进新的 NFC 服务，NFC 论坛继续通过降低市场准入壁垒的新规范来清除商业化的道路，这会促进新的想法并将开发人员聚集到一起测试其产品的互操作性。

每个解决方案开发人员都应该有一个 NFC 的策略。10 年前，许多解决方案开发人员慢慢看到移动设备会如何影响他们的产品及其使用。用户确实受益于首先推出移动版本或应用程序，从而更容易，更方便，访问更多信息或新功能。NFC 也提供类似的机会。埃文斯数据公司（Evans Data Corporation）最近发布的“移动开发调查”（Mobile Development Survey）指出，目前有超过 31% 的移动开发商正在其移动应用中支持 NFC。尽管这是一个令人鼓舞的迹象，另外的 69% 应当开始评估它们如何利用 NFC 更好地满足客户的需求，促进他们的业务战略。

商业需要探索 NFC 如何驱动新的增长。第一个移动“解决方案”几乎不比适配到小手机屏幕上以合适显示的传统网站多什么。然而，在几年之内移动自身成为市场板块，引发了一波仍在上升的创新潮流。NFC 具有相同的变革能力——事实上，由于 NFC 超越了移动设备，因此可以实现互联网与日常设备和对象之间的链接，从家用电器到商店的标牌都可以。随着这些日常物品中嵌入更多的智能，NFC 有能力以新的方式使这种智能工作用于内部业务和外部客户应用程序。

例如，法国的一家公司开发了一种支持 NFC 的技术，可以监控分销渠道的葡萄酒出货温度，以确保葡萄酒的出处和品质。每箱葡萄酒都配有电池供电的射频识别（RFID）[3] 温度传感器。在分配周期的每个步骤中，可以使用支持 NFC 的设备检查葡萄酒的温度和真实性。如果没有 NFC，这种解决方案将是昂贵的和不切实际的。NFC 的商业优势有很多。拥有移动办公人员的公司，如旅行服务人员，可以用 NFC 更好地跟踪和指导他们的行动。寻求与有价值的消费者建立 1∶1 营销关系的市场人员、广告商和零售商可以在他们具有最大影响的一点——销售——上利用 NFC，提供个性化的服务。

第 12 单元

课文 A　数字信号处理

数字信号处理（DSP）是将信号以数字方式表示并处理的理论和技术，如通过计算机执行一系列信号处理操作。以这种方法处理的信号是表示连续变量样本的一系列数字，如时间域、空间域或频域。

数字信号处理和模拟信号处理是信号处理的子领域。DSP 的主要应用是音频和语音信号处理、声呐、雷达和其他传感器阵列处理、频谱估计、统计信号处理、数字图像处理、电信信号处理、控制系统、生物医学工程、地震数据处理等。

数字信号处理包括线性或非线性操作。非线性信号处理与非线性系统辨识密切相关而且可以在时间、空间和时空领域实现。

数字信号处理计算的应用程序在许多应用程序中有许多优于模拟处理的优点。例如，传输中的错误检测和校正以及数据压缩。DSP 适用于流数据和静态（存储的）数据（见图 12-1）。

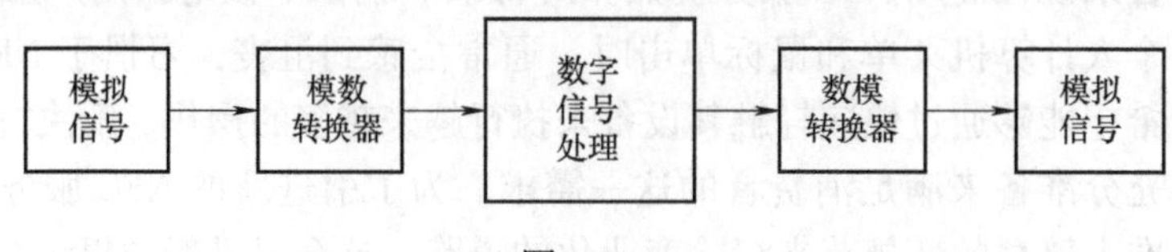

图　12-1

1. 信号采样

计算机使用的增加导致越来越多的数字信号处理的使用和需求增加，要数字化地分析和操作模拟信号就必须使用模数转换器。采样通常分两个阶段进行：离散化和量化。离散化意味着信号被分成相等的时间间隔，并且每个间隔由单个测量的幅度值表示。量化意味着每个幅度测量值都有一个来自有限集合的值。将实数舍入到整数就是一个例子。

奈奎斯特-香农采样定理表明，如果采样频率大于信号最高频率的两倍，则可以从其采样中精确地重建信号，但这需要无限数量的量化级别。实际上，采样频率通常明显高于信号有限带宽所需的两倍。

理论的 DSP 分析和推导通常在离散时间信号模型上执行，没有幅度失真（量化误差），通过抽象的样本处理进行“创建”。数字方法需要量化信号，如由模拟-数字转换器（ADC）[1]产生的信号。处理的结果可能是频谱也可能是一组统计。不过它通常是由数模转换器（DAC）转换成模拟形式的另一个量化信号。

2. 领域

在 DSP 中，工程师通常在以下领域之一研究数字信号：时域（一维信号）、空间域（多维信号）、频域和小波域。他们基于哪个域最好地表示信号的本质特性作为经验假设（或通过尝试不同的可能性），来选择在哪个域处理信号。一系列来自测量装置的样本产生时间或空间域的表达，而离散傅里叶变换产生信息频域，即频谱。

3. 时间和空间域

时间或空间域中最常见的处理方法是通过称为滤波的方法增强输入信号。数字滤波通常由围绕输入或输出信号的当前样本周围的多个周围样本的一些线性变换组成。有各种方法来表征过滤器，例如：

1）“线性”滤波器是对输入的样本进行线性变换；其他滤波器则是“非线性”的。线性滤波器满足叠加条件，即如果输入的是不同信号的加权线性组合，则输出是相应输出信号的类似加权的线性组合。

2）“因果”滤波器仅使用输入或输出信号的先前样本；而“非因果”滤波器使用未来的输入样本。非因果滤波器通常可以通过添加延迟变成因果滤波器。

3）“时不变”滤波器随时间推移具有不变的属性；其他滤波器如自适应滤波器随时间变化。

4）“稳定”滤波器产生随时间收敛到恒定值的输出，或者在有限的间隔内保持有界。“不稳定”滤波器在有界甚至零输入时可以产生增长到无边界的输出。

5）“有限脉冲响应”（FIR）滤波器仅使用输入信号，而“无限脉冲响应”滤波器（IIR）既要使用输入信号也要使用输出信号的先前采样。FIR 滤波器总是稳定的，而 IIR 滤波器可能不稳定。

滤波器可以由框图表示，然后可以使用它来导出用硬件指令实现滤波器的样本处理算法。滤波器也可以被描述为差分方程、零点和极点的集合或者脉冲响应、阶跃响应，如果它是 FIR 滤波器。

对任何给定输入，可以通过将输入信号与脉冲响应进行卷积来计算线性数字滤波器的输出。

4. 频率域

信号通过傅里叶变换从时域或空域转换到频域。傅里叶变换将信号信息转换成每个频率的幅度和相位分量。通常，傅里叶变换被转换为功率谱，这是每个频率分量大小的平方。

在频域信号分析中最常见的是信号特性的分析。工程师可以研究频谱，以确定输入信号中哪些频率存在哪些频率丢失。

除了频率信息之外，通常还需要相位信息，这可以从傅里叶变换获得。在一些应用中，相位如何随频率而发生变化需要仔细考虑。

特别是在非实时工作中，也可以应用滤波器转换到频域再转换回时域来实现过滤。这是一个快速的 O（nlogn）操作，而且基本上可以给出任何滤波器形状，包括对矩形滤波器的非常相似的形状。

还有一些常用的频域的转换。例如，倒频谱通过傅里叶变换将信号转换为频域，取对数，然后应用另一傅里叶变换。这强调了原频谱的谐波结构。频域分析也被称作频谱分析。

5. Z 平面分析

数字滤波器采用 IIR 和 FIR 类型。FIR 滤波器具有许多优点，但是在计算上要求更高。IIR 滤波器具有可能谐振的反馈环路，所以当用某些输入信号刺激时，FIR 滤波器总是稳定的。Z 变换提供了一个能够分析数字 IIR 滤波器的潜在的稳定性问题的工具。它类似于拉普拉斯变换，用于设计模拟 IIR 滤波器。

6. 小波

在数值分析和功能分析中，离散小波变换（DWT）是小波离散采样的任何小波变换。与其他小波变换一样，离散小波变换具有超过傅里叶变换的一个巨大优点是时间分辨率：它捕获频率和位置信息。

7. 应用

DSP 的主要应用是音频信号处理、音频压缩、数字图像处理、视频压缩、语音处理、语音识别、数字通信、数字合成器、雷达、声呐、财务信号处理、地震学和生物医学。具体例子如：数字手机中的语音压缩和传输、高保真音响和声音增强应用中的室内声音校正、天气预报、经济预测、地震数据处理、工业过程的分析和控制、医学成像如 CAT 扫描和 MRI、MP3 压缩、计算机图形、图像处理、高保真扬声器分频器和均衡，以及用于电吉他放大器的音频效果。

8. 实现

长期以来，DSP 算法一直在通用的计算机和数字信号处理器上运行。DSP 算法也在专用集成电路（ASICs）[2] 等专用硬件上实现。数字信号处理的附加技术包括更强大的通用微处理器、现场可编程门阵列（FPGAs）[3]、数字信号控制器（主要用于工业应用，如电机控制）和流处理器。

根据应用的要求，数字信号处理任务可以在一般计算机上实现。通常当处理要求不是实时的时候，为了节约经济，可以通过现有的一般用途的计算机处理数据文件中存在的信号数据（输入或输出）。除了使用 DSP 数学技术（如 FFT）之外，与其他数据处理基本上没有任何区别，并且采样数据通常被认为在时间上或空间上均匀采样。例如：使用 Photoshop 等软件处理数码照片。

然而，当应用需求是实时的时候，DSP 通常使用专门的微处理器（如 DSP56000，

TMS320 或 SHARC）来实现。虽然一些更强大的版本使用浮点数，但是这些经常使用定点算法处理数据。对于更快的应用，可能会使用 FPGA。从 2007 年开始，DSP 的多核实现已经开始从包括飞思卡尔[4]和流处理器公司在内的公司涌现。为了更快地应用程序，ASIC 可能会被专门设计。对于缓慢的应用，传统的较慢处理器（如微控制器）应该是足够的。而且，越来越多的 DSP 应用采用具有多核处理器的强大个人计算机部署于嵌入式系统。

课文 B　数字信号处理器

数字信号处理器（DSP）是专门的微处理器，或者可以说是一个集成电路封装块（SIP）。其构架针对数字信号处理器的工作需要进行优化。

DSP 的目的通常是进行测量、滤波或压缩真实世界的连续的模拟信号。大多数通用微处理器也可以进行数字信号的处理算法，但是专门的数字信号处理器通常功耗低，因为功耗有限制，所以它们更适合在电话这类便携设备上使用。DSP 使用特殊的存储构架来同时获取多个数据和指令。

1. 概述

数字信号处理算法通常需要对一系列数据样本进行快速而又反复的大量数学运算。信号（可能来自音频或视频传感器）不断地从模拟信号转为数字信号，之后进行数字运算，又返还成模拟信号。许多数字信号处理器的应用在时延上有限制，就是说为了使系统工作，数字信号处理操作必须在固定时间内完成，因此延迟（或分批处理）是不可行的。

大多数通用微处理器和操作系统可以顺利地运行数字信号处理算法，但由于功耗限制，并不适合用在像手机和掌上电脑这些便携设备上。然而，专门的数字信号处理器倾向于提供低成本的解决方案，拥有更好的性能、更低的延迟，并且不需要专门的冷却或是大电池。

DSP 的构架特意为数字信号处理进行了优化。大多数处理器也支持应用处理器、微控制器等特性，因为在很少的情况下系统只有信号处理这一单一任务。一些用于优化 DSP 算法的有用特性在下文概述。

2. 构架

（1）软件

按照通用处理器的标准，数字信号处理器的指令集通常是相当不符合规则的。传统指令集是由更多一般的指令组成，以便处理器可以执行更多种类的运算。DSP 则针对数字信号处理器计算中频繁出现的数学运算进行优化。传统的和 DSP 优化的指令集的处理器都能执行任意的运算，然而需要多条 ARM 或 x86 指令的运算对于优化过的 DSP 指令集可能只需要一条指令。

软件构架的一个含义是人工优化过的汇编代码流程通常打包成库以重用，而不是依靠高级编译器技术来解决基本算法。即使使用最新的编译器进行优化，也不如人工优化汇编代码更高效。DSP 运算涉及的许多常见算法由人工编写，以充分利用框架优化。

1）指令集。

- 乘积累加运算［MACs，包括融合乘加（FMA）］；
- 广泛用于各种矩阵运算；
- 卷积滤波；
- 点积；

- 多项式求值；
- 基本的 DSP 算法高度依赖于乘积累加的性能；
- FIR 滤波器；
- 快速傅里叶变换；
- 提升并行化的指令；
- 单指令流多数据流；
- 超长指令字；
- 超级标量架构；
- 用于 FFT 交叉参考的环形缓冲区中的模寻址和位反转寻址专用指令；
- DSP 有时使用时间稳定的编码来简化硬件并提高编码效率；
- 多个运算单元可能需要存储器的构架来支持每个指令循环的存取；
- 特殊的循环控制，譬如在一个非常紧密的循环中有构架的支持，执行几个指令就无须用于取指令或是退出测试的系统开销。

2）数据运算指令。

- 使用饱和算术运算中溢出的值将会累积至存储器可容纳的最大值（或是最小值），而不是进行环绕式处理（在环绕方式中，最大值 +1 溢出变为最小值，正如在许多通用 CPU 中那样；使用饱和算术时它仍保持在最大值）。有些情况下可以使用不同的黏滞位运算模式；
- 定点运算通常用来加速算术处理；
- 单循环操作增强流水线操作的益处。

3）程序流。

- 浮点单元直接融进数据路径当中；
- 流水线架构；
- 高度并行的乘法累加运算（MAC 单元）；
- 硬件控制循环以减少或消除循环操作需要的系统开销。

（2）硬件

1）内存构架。数字信号处理器通常针对流数据进行优化，并使用能同时获取多数据和指令的内存架构。譬如哈佛架构或修正的冯·诺依曼结构，其使用独立的程序和数据存储器（甚至有时可以并发访问总线上的数据）。

2）寻址和虚拟内存。数字信号处理器经常使用多任务操作系统，但不支持虚拟内存和内存保护。使用虚拟内存的操作系统需要更多的时间用于程序间的内容切换，这增加了延迟。

- 硬件模寻址；
- 允许执行循环缓冲而不用检测换行；
- 位反转寻址，一个特殊寻址模式；
- 对计算快速傅里叶变换有用；
- 无须内存管理单元；
- 内存地址计算单元。

3. 历史

在下述讨论的独立 DSP 出现之前，大多数 DSP 应用都是使用位片处理器实现的。AMD2901 位片芯片及其系列组件是非常受欢迎的选择。有参考 AMD 的设计，但是特定设计的细节往往是由具体应用决定的。这些位片架构有时会包括一个外设的乘法芯片。这些乘法芯片的例子是来自 TRW 的包含 TDC1008 和 TDC1010 的系列芯片，系列中的一些包含累加器，提供乘积累加（MAC）功能。

在 1976 年，理查德・威金斯向德州仪器达拉斯研究中心的保罗・布里德洛夫，拉里・布兰丁汉和吉恩・弗朗茨提出了 Speak & Spell 的概念。两年之后的 1978 年他们研发出第一台 Speak & Spell，其技术核心是 TMS5100，业界首个 DSP。它还第一个使用线性预测编码来执行语言合成的芯片。

在 1978 年，英特尔发布了“模拟信号处理器”2920 处理器。它包含一组带有一个内部信号处理器的片上 ADC/DAC，但由于它不含硬件乘法器因此在市场上的销售并不成功。在 1979 年，AMI 发布了 S2811，它被设计成微处理器的外围设备，必须由主处理器初始化后才能工作。因此 S2811 在市场上销售也不成功。

1980 年第一批独立完整的 DSP——NEC μPD7720 和 AT&T DSP1 在 1980 年国际固态电路会议上出展。这两种处理器都是受到公共交换电话网[1]的研究的启发。

Altamira DX-1 是另一个早期的 DSP，它使用一组带有延迟分支和分支预测机制的四整数组的流水线。

另一种由得克萨斯仪器公司[2]生产的数字信号处理器 TMS32010 在 1983 年诞生，被证明取得了更大的成功。它是建立在哈佛架构上的，因此也有独立的指令和数据存储器。它已经有一个特殊的指令集，具有加载-累加，乘-累加的指令。它可以在 16 位上工作，并进行乘法运算，只要 390ns。如今，该公司已成为通用 DSP 市场的龙头。

约 5 年后，第二代 DSP 开始传播。它们有三个存储器用来同时存储两个运算符，并且包含硬件以加速紧环，它们还具有能够进行环路寻址的寻址单元。其中一些操作在 24 位变量上，典型的模型进行 MAC 运算只需要 21ns。这一代成员有 AT&T DSP16A 或摩托罗拉 56000。

第三代 DSP 的主要的改进在于针对特定应用的单元以及数据路径中指令的出现，有时作为协处理器。这些单元格允许直接硬件加速，进行非常具体但复杂的数学问题，如傅里叶变换或矩阵运算。一些芯片，如摩托罗拉 MC68356，甚至包括多个处理器内核以并行工作。1995 年以后的其他数字信号处理器包括 TI TMS320C541 和 TMS 320C80。

第四代 DSP 的特点是指令集和指令编码/解码的变化。增加了单指令多数据流（SIMD）[3]扩展，出现了 VLIW 和超标量架构。像以往一样，时钟速度又提高了，3nsMAC 运算成为可能。

4. 现代 DSP

现代信号处理器可以带来更高的性能，这部分归于技术和架构的进步，如更低的设计规则，快速访问二级缓存，直接内存存取电路，以及更宽的总线系统。不是所有的 DSP 都规定了相同的速度，并且还存在多种信号处理器，每一款都适用于特定的任务，价格从约 1.5 美元到 300 美元不等。得克萨斯仪器公司生产的 C6000 系列的 DSP 具有 1.2GHz 频率并执行分立的指令和数据缓存。它们还拥有 8MB[⊖]二级缓存和 64 个 EDMA 通道。最好的模型能够每秒高达 8000MIPS，使用 VLIW[4]（超长指令字），在每个时钟周期执行 8 次运算，并且能与广泛的外设和各种总线（PCI/串行/……）兼容。每一个 TMS320C6474 芯片都有三个这样的 DSP，最新一代的 C6000 芯片支持浮点和定点处理。

飞思卡尔生产的多核 DSP 系列——MSC81xx。MSC81xx 基于星核架构处理器，最新的 MSC8144DSP 组合了四个可编程 SC3400 StarCore DSP 内核。每个 SC3400 StarCore DSP 内核的时钟速度为 1GHz。

XMOS 生产了一种适用于 DSP 操作的多核多线程处理器，它们的速度范围为400～1600 MIPS。处理器具有多线程架构，每个内核最多可以提供8 个实时线程，意味着4 个内核设备最多可以支持 32 个实时线程。线程之间通过缓冲通道通信，速度能够高到 80Mbit/s。这些器件可用 C 语言轻松编程，旨在缩小传统微控制器和 FPGA 之间的差距。

CEVA 公司生产和许可了三个不同的 DSP 系列。也许最知名和最广泛部署的是 CEVA-TeakLite DSP 系列，这是一种经典的基于内存的架构，具有 16 位或 32 位字宽的单 MAC 和双 MAC。CEVA-X DSP 系列提供了 VLIW 和 SIMD 架构的结合，该系列的不同成员提供 2 位或 4 位的 16 位 MAC。CECA-XC DSP 系列针对软件定义无线电（SDR）调制解调器设计，并利用具有 32 位和 16 位 MAC 的 VLIW 和矢量架构的独特组合。

Analog Devices 公司生产的基于 SHARC 的 DSP 的性能范围为 66MHz/198 MFLOPS（每秒百万浮点运算）～400MHz/2400 MFLOPS。一些型号支持多个乘法器和算术逻辑单元，SIMD 指令和音频特定的处理部件以及外设。Blackfin 家族嵌入式数字信号处理器结合了通用处理器的特点，结果是这些处理器可以运行简单的操作系统，像 μCLinux，velOSity 和 Nucleus RTOS，进行实时数据操作。

⊖ $1MB = 2^{20}B = 1024KB = 1048576bit$

Answers to Questions for Discussion (for reference)

Unit 1 Text A

1. What's the main idea of this text?

This passage gives a detailed and brief account of the history of telecommunication and those who contributed much to the development of telecommunication systems.

2. What do you learn about the first commercial telephone?

The first commercial telephone services were set up in 1878 and 1879 on both sides of the Atlantic in the cities of New Haven, Connecticut, and London, England. The technology grew quickly from this point, with inter-city lines being built and telephone exchanges in every major city of the United States by the mid-1880s. The First transcontinental telephone call occurred on January 25, 1915. Despite this, transatlantic voice communication remained impossible for customers until January 7, 1927 when a connection was established using radio.

3. How did Samuel Morse contribute to the development of telegraph?

Samuel Morse developed a version of the electrical telegraph which he demonstrated on September 2, 1837. Alfred Vail saw this demonstration and joined Morse to develop the register—a telegraph terminal that integrated a logging device for recording messages to paper tape. Morse's most important technical contribution to this telegraph was the simple and highly efficient Morse Code, co-developed with Vail, which was an important advance over Wheatstone's more complicated and expensive system, and required just two wires. The communications efficiency of the Morse Code preceded that of the Huffman code in digital communications by over 100 years, but Morse and Vail developed the code purely empirically, with shorter codes for more frequent letter.

Unit 1 Text B

1. How did Heinrich learn Greek and Latin? Why?

He received tutoring in Greek and Latin. Because Dr. Lange's school did not teach Greek and Latin—the classics needed for university entry. At the age of 15, Heinrich left Dr. Lange's school to be educated at home and he had decided that perhaps he would like to go to university after all, he learned to prepare him for the exams.

2. What kind of routine did Hertz fall into at the University of Berlin?

Hertz started work on the problem and quickly fell into a pleasant routine: Attending a lecture each morning in either analytical dynamics or electricity & magnetism, carrying out experiments in the laboratory until 4pm, then reading, calculating, and thinking in the evening.

3. How did he prove Maxwell's theory?

Over the next three years, in a series of brilliant experiments, Hertz fully verified Maxwell's theory. He proved beyond doubt that his apparatus was producing electromagnetic waves, demonstrating that the energy radiating from his electrical oscillators could be reflected, refracted, produce interference patterns, and produce standing waves just like light.

Unit 2 Text A

1. What are the basic elements of wireless technologies?

- Spectrum, or the range of frequencies in which the network operates;
- Transmission speeds supported;
- Underlying transmission mechanism, such as frequency division multiple access (FDMA), time division multiple access (TDMA), or code division multiple access (CDMA);
- Architectural implementation, such as enterprise based (or in-building), fixed, or mobile.

2. What components is a typical cellular telecommunications network made up of?

A typical cellular telecommunications network consists of the following components:

- public switched telephone Network (PSTN);
- mobile switching center (MSC);
- base station (BS);
- radio access network (RAN);
- home location register (HLR);
- visitor location register (VLR);
- authentication center (AUC).

3. What is MSC? What does it serves as in the mobile wireless network infrastructure?

MSC, namely, mobile switching center usually, is located at the Mobile Telephone Switching Office, which is part of the mobile wireless network infrastructure.

The major functions of MSC include:

- Switching voice traffic from the wireless network to the PSTN; it switches to another MSC;
- Providing telephony switching services and controls calls between telephone and data systems;
- Providing the mobility functions for the network and serves as the hub for up to as many as 100 BSs.

4. What is HLR and VLR? What are their functions?

Home location register (HLR) is a database that contains information about subscribers to a mobile network that is maintained by a particular service provider. And visitor locations register (VLR) is a database that is maintained by an MSC, to store temporary information about subscribers who roam into the coverage area of that MSC.

The HLR stores "permanent" subscriber information (rather than temporary subscriber data, which a VLR manages), including the service profile, location information, and activity status of the mobile user. The VLR, which is usually part of an MSC, communicates with the HLR of the roaming subscriber to request data, and to maintain information about the subscriber's current location in the network

Unit 2 Text B

1. What is 3G?

The "G" stands for Generation. 3G made its debut in 2001, and included multi-media support along with a peak transfer rate of at least 200 kilobits per second. 3G set the standards for most of the wireless technology we have come to know and love. Web browsing, e-mail, video downloading, picture sharing and other smartphone technology were introduced in the third generation. 3G should be capable of handling around 2 Megabits per second.

2. What exactly does LTE mean?

LTE actually stands for "long term evolution", and its full name is 3GPP LTE, with the 3GPP standing for the 3rd Generation Partnership Project, which has been developing the technology's release documents. It's based on GSM/EDGE and UMTS/HSPA network technologies, and provides an increase to both capacity and speed using new techniques for modulation. The new architecture of LTE technology means lower operating costs along with greater overall data and voice capacity.

3. What does 5G stand for?

5G stands for the fifth generation of wireless technologies and it will be faster than 4G. 5G is going to be fast. It is supposed to be fast enough for everyone everything and the IoT (the Internet of things). 5G usage goes way beyond your smartphone and devices. This will be what drives your cars, it will allow machines to communicate and pretty much anything else that will benefit from being connected.

Unit 3 Text A

1. What kind of online conveniences can we enjoy through use of websites on the World Wide Web?

Nowadays we can enjoy a large quantity of online conveniences by using websites on the World Wide Web, which include:

- Communication via e-mail and VoIP;
- Sharing of information such as text, images, sounds and videos;
- Storage of information;
- Streaming television programmes, films, videos, sounds and music;
- Playing online games;
- Shopping;
- Social networking;
- Banking.

2. Can you list some disadvantages of e-mails? What are they?

Here are main disadvantages of e-mails:

1) The recipient can only receive the e-mail if they are connected to the Internet.

2) E-mails can sometimes contain viruses in the form of attachments.

3) Spam e-mails can be a problem. So can phishing e-mails, which are designed to trick people into giving away personal information.

4) Because e-mails can be delivered to Internet-connected digital devices anywhere, they can be hard to get away from.

3. What are major advantages of video conferencing?

Video conferencing bears many advantages, of which the most important ones are as follows:

• Seeing as well as hearing the other person.

• Showing others what is going on around us.

• Reducing time to travel to see and speak with someone. This has even greater benefits if the other person is on the other side of the world.

• Saving money, in travel costs.

• The ability to video conference several people in different locations, at the same time.

4. In what aspects do you think Internet will further influence us?

开放性问题，答案略。

Unit 3 Text B

1. How many forms of Internet-based communications are introduced in the essay? What are they?

Six. Instant messaging, Internet telephone & VoIP, e-mail, IRC (Internet relay chat), video-conferencing, SMS (short message service) and wireless Communications.

2. What are the problems and issues associated with IM?

Some problems and issues associated with IM include spim and virus propagation.

3. What are the advantages of video-conferencing?

Video-conferencing is a conference between two or more participants at different sites by using computer networks to transmit audio and video data. It has become a cost efficient way to provide distance learning, guest speakers, and multi-school collaboration projects. Many feel that video-conferencing provides a visual connection and interaction that cannot be achieved with standard IM or e-mail communications.

Unit 4 Text A

1. Why are metal wires inferior to fiber optics?

While electricity travels from one end to the far end, the signal strength degrades as the energy experiences a type of electrical friction called impedance, which results in the signal decaying over distance and the wire becoming warmer, which can cause some problems. On the other hand, fiber optics are characterized by less signal degradation, more untapped overhead and easier upgrades.

2. Why are fiber optics, compared with metal wires, much easier to upgrade?

Fiber optics have the combination of substantially greater distance between network nodes and substations and more untapped overhead, making upgrades less of a hassle for network carriers, which, in turn, means less fees that have to be passed on to the consumers.

3. Why do those offer fiber optic networks of DSL tend to have their fiber close to the homes of their customers as possible as they can?

Because they know that fiber optics are the most cost effective solution, and they know that by putting fiber close to the homes and businesses that they serve that they stand a very good chance of making the transition to an all-fiber network that much easier.

4. What do you think of the discovery made by the team at Trinity?

In spite of the fact that this discovery is still in its infancy. In a sense, it is a breakthrough for the world of physics and science alike. It's just a matter of time before data communication companies come knocking on the doors of Trinity College Dublin, wanting to bring this new tech to the forefront of mass communications.

Unit 4 Text B

1. What is the advantage of fiber optic cables over copper cables?

Fiber optic cables are able to carry much more data than copper cables, especially over long distances.

2. How do you understand "fiber is future-proof"?

Fiber is often said to be "future-proof" because the data rate of the connection is usually limited by the terminal equipment rather than the fiber, permitting substantial speed improvements by equipment upgrades before the fiber itself must be upgraded.

3. What is the difference between FTTH and FTTB?

An apartment building may provide an example of the distinction between FTTH and FTTB. If a fiber is run to a panel inside each subscriber's apartment unit, it is FTTH. If instead the fiber goes only as far as the apartment building's shared electrical room (either only to the ground floor or to each floor), it is FTTB.

Unit 5 Text A

1. What does the Internet of things mean?

The Internet of things (stylized Internet of Things or IoT), a term coined by Peter T. Lewis in September 1985, which is defined as "the infrastructure of the information society" by the Global Standards Initiative on Internet of Things (IoT-GSI) in 2013, refers to the Internetworking of physical devices, vehicles (also referred to as "connected devices" and "smart devices"), buildings, and other items—embedded with electronics, software, sensors, actuators, and network connectivity that enable these objects to collect and exchange data.

2. What do you think of the future of IoT?

IoT, it seems to me, will find wide use in our daily life in the future. According to Gartner, Inc. (a technology research and advisory corporation), there will be nearly 20. 8 billion devices on the Internet of things by 2020. ABI research estimates that more than 30 billion devices will be wirelessly connected to the Internet of things by 2020. As per a 2014 survey and study done by Pew Research Internet Project, a large majority of the technology experts and engaged Internet users who responded—83 percent—agreed with the notion that the Internet/Cloud of Things, embedded and wearable computing (and the corresponding dynamic systems) will have widespread and beneficial effects by 2025.

3. How many types can medium-range wireless fall into? What are they?

Medium-range wireless, generally speaking, can fall into two types:

1) HaLow—Variant of the WiFi standard that provides extended range for low-power communication at a lower data rate.

2) LTE-Advanced—High-speed communication specification for mobile networks, which provides enhancements to the LTE standard with extended coverage, higher throughput, and lower latency.

Unit 5 Text B

1. What are connected cars?

"Connected vehicles" are cars that access, consume, create, enrich, direct, and share digital information between businesses, people, organizations, infrastructures, and things. Those "things" include other vehicles, which is where the Internet of things becomes the Internet of cars.

2. What are the benefits and opportunities of connected cars?

Reduced accident rates, increased productivity, improved traffic flow, lowered emissions, extended utility for EVs, new entertainment options, and new marketing and commerce experiences.

3. What is the demand of U. S. vehicle owners according to the analyses Gartner conducted?

Almost half (46%) are interested in safely accessing mobile applications inside the vehicle. These applications include receiving on-demand wireless map or software updates, finding available parking spots, and conducting local searches; nearly 40% would also opt for remote diagnostic capabilities that alert them when parts need replacement.

More than one-third are interested in a self-driving, autonomous vehicle.

Thirty percent are likely to opt for a vehicle that allows them to tether their smartphone to get Internet connection there.

Unit 6 Text A

1. Is there any disadvantages in conventional data-encryption systems? What is it?

Conventional data-encryption systems rely on the exchange of a secret "key" —in binary 0s and 1s—to encrypt and decrypt information. But the security of such a communication channel can be undermined if a hacker "eavesdrops" on this key during transmission.

2. How can QKD prevent hackers from eavesdropping during transmission?

Quantum communications use a technology called quantum key distribution (QKD), which harnesses the subatomic properties of photons to remove this weakest link of the current system, which, therefore, allows a user to send a pulse of photons that are placed in specific quantum states that characterize the cryptographic key. If anyone tries to intercept the key, the act of eavesdropping intrinsically alters its quantum state—alerting users to a security breach.

3. What is the disadvantage of dark fibers? How can it be avoided?

The problem with dark fibers is that they are not always available and can be prohibitively expensive. One way to sidestep the problem is to piggyback the photon streams onto the "lit" fibers that transmit conventional telecommunications data.

4. Do you think it is possible that QKD will be employed in "real life"?

Yes. As long as dark fibers can be replaced with optical fibers, it can be used in "real life" . In fact, Shields' research has shown the multiplexing of strong classical signals with quantum signals in the same fiber for the first time, which indicates it is possible to remove the need for dark fibers.

Unit 6 Text B

1. What project has received close attention in scientific and security circles for the past five years?

China sent the world's first quantum-communications satellite into orbit from a launch center in Inner Mongolia about 1: 40 a. m. Tuesday.

2. How do you understand "quantum encryption is secure"?

Quantum encryption is secure against any kind of computing power because information encoded in a quantum particle is destroyed as soon as it is measured.

3. Why is quantum encryption not foolproof?

Quantum encryption isn't foolproof. It's possible for hackers to trick an incautious recipient by shining an intense laser into a quantum receptor.

Unit 7　Text A

1. What is stored program control?

Stored program control (SPC) is a telecommunications technology used for telephone exchanges controlled by a computer program stored in the memory of the switching system. SPC was the enabling technology of electronic switching systems (ESS) developed in the Bell System in the 1950s.

2. What main feature does SPC bear?

The main feature of SPC is one or multiple digital processing units (stored program computers) that execute a set of computer instructions (program) stored in the memory of the system by which telephone connections are established, maintained, and terminated in associated electronic circuitry.

3. How many categories does SPC fall into? What are they?

SPC falls into two categories: centralized SPC and distributed SPC. More specifically, stored program control implementations may be organized into centralized and distributed approaches. Early electronic switching systems (ESS) developed in the 1960s and 1970s almost invariably used centralized control. Although many present day exchange design continue to use centralized SPC, with advent of low cost powerful microprocessors and VLSI chips such as programmable logic array (PLA) and programmable logic controllers (PLC), distributed SPC became widespread by the early 21st century.

4. What is standby mode?

Standby mode of operation is usually referred to as the simplest of a dual-processor configuration. Normally, one processor is in standby mode. The standby processor is brought online only when the active processor fails. An important requirement of this configuration is ability of standby processor to reconstitute the state of exchange system when it takes over the control; means to determine which of the subscriber lines or trunks are in use.

Unit 7　Text B

1. What is the feature of packet switching?

Packet switching features delivery of variable bit rate data streams, realized as sequences of packets, over a computer network which allocates transmission resources as needed using statistical multiplexing or dynamic bandwidth allocation techniques. As they traverse network nodes, such as switches and routers, packets are received, buffered, queued, and transmitted (stored and forwarded), resulting in variable latency and throughput depending on the link capacity and the traffic load on the network.

2. What does Baran report in 1962 focus on?

Report P-2626 described a general architecture for a large-scale, distributed, survivable communications network. The work focuses on three key ideas: Use of a decentralized network with multiple paths between any two points, dividing user messages into message blocks, later called packets, and delivery of these messages by store and forward switching.

3. List some examples of connectionless protocols.

Examples of connectionless protocols are Ethernet, Internet Protocol (IP), and the User Datagram Protocol (UDP). Connection-oriented protocols include X.25, Frame Relay, Multiprotocol Label Switching (MPLS), and the Transmission Control Protocol (TCP).

Unit 8 Text A

1. What does information security refer to?

Information security, InfoSec for short, refers to the practice of preventing unauthorized access, use, disclosure, disruption, modification, inspection, recording or destruction of information. It is a general term that can be used regardless of the form the data may take (e.g. electronic, physical).

2. What is IT security?

Sometimes referred to as computer security, information technology security (IT security) is information security applied to technology (most often some form of computer system). Here a computer does not necessarily mean a home desktop. A computer is any device with a processor and some memory. Such devices can range from non-networked standalone devices as simple as calculators, to networked mobile computing devices such as smartphones and tablet computers.

3. What are the most common threats to information security?

Some of the most common threats today are software attacks, theft of intellectual property, identity theft, theft of equipment or information, sabotage, and information extortion. Most people have experienced software attacks of some sort. Viruses, worms, phishing attacks, and Trojan horses are a few common examples of software attacks.

Unit 8 Text B

1. What should organizations do in the drive to become more cyber resilient?

In the drive to become more cyber resilient, organizations need to extend their risk management focus from pure information confidentiality, integrity and availability to include risks such as those to reputation and customer channels, and recognize the unintended consequences from activity in cyberspace.

2. What should Chief Information Security Officers (CISOs) do when preparing to embrace the increasingly complex IoT?

Chief Information Security Officers (CISOs) should be proactive in preparing the organization for the inevitable by ensuring that apps developed in-house follow the testing steps in a recognized systems development lifecycle approach. They should also be managing user devices in line with existing asset management policies and processes, incorporating user devices into existing standards for access.

3. How can government departments, regulators, senior business managers and information security professionals better understand the true nature of cyber threats?

By adopting a realistic, broad-based, collaborative approach to cyber security and resilience, government departments, regulators, senior business managers and information security professionals will better understand the true nature of cyber threats and how to respond quickly and appropriately.

Unit 9 Text A

1. In this passage, what does multiplexing mean?

Multiplexing is the sharing of a communications channel through local combining of signals at a common point. Multiplexing commonly falls into two types: frequency-division multiplexing and time-division multiplexing.

2. Which two kinds of multiplexing are frequently used in order for multiple users to share the communications channel?

Both frequency-division multiplexing (FDM) and time-division multiplexing (TDM) are frequently used to enable users share the channel. The former means the available bandwidth of a communications channel is shared among multiple users by frequency translating, or modulating, each of the individual users onto a different carrier frequency and the latter refers to the process in which multiplexing is conducted through the interleaving of time segments from different signals onto a single transmission path.

3. What are the three schemes devised for efficient sharing of a single channel?

The three schemes are called frequency-division multiple access (FDMA), time-division multiple access (TDMA), and code-division multiple access (CDMA). These techniques can be used alone or together in telephone systems.

Unit 9 Text B

1. What is the primary advantage of OFDM over single-carrier schemes?

The primary advantage of OFDM over single-carrier schemes is its ability to cope with severe channel conditions (for example, attenuation of high frequencies in a long copper wire, narrowband interference and frequency-selective fading due to multipath) without complex equalization filters.

2. List at least four advantages of OFDM.

- High spectral efficiency as compared to other double sideband modulation schemes, spread spectrum, etc.;
- Can easily adapt to severe channel conditions without complex time-domain equalization;
- Robust against narrow-band co-channel interference;
- Robust against intersymbol interference (ISI) and fading caused by multipath propagation;
- Efficient implementation using fast Fourier transform (FFT);
- Low sensitivity to time synchronization errors;
- Tuned sub-channel receiver filters are not required (unlike conventional FDM);
- Facilitates single frequency networks (SFNs) (i. e. transmitter macrodiversity).

3. What is the additional constraint of OFDM?

Conceptually, OFDM is a specialized FDM, the additional constraint being that all carrier signals are orthogonal to one another.

Unit 10　Text A

1. What are the major reasons for application of modulation to information signals?

The three major reasons are as follows：

First, modulation methods have to be applied to the information signals in order for "frequency to translate" the signals into the range of frequencies that are permitted by the channel.

Second, each user's information signal needs to be modulated onto an assigned carrier of a specific frequency to prevent mutual interference.

Third, the application modulation to information signals contributes to the translation of the voice frequency to a higher frequency, which allows the use of a much smaller antenna.

2. What are the frequently employed methods of modulating analog signals?

Two commonly used methods are used to modulate analog signals. One technique, called amplitude modulation, varies the amplitude of a fixed-frequency carrier wave in proportion to the information signal. The other technique, called frequency modulation, varies the frequency of a fixed-amplitude carrier wave in proportion to the information signal.

3. Do you know any other advanced forms of digital modulation?

There are another two advanced methods of digital modulation: quadrature amplitude modulation (QAM) and trellis-coded modulation (TCM). QAM signals actually transmit two modulated signals in phase quadrature (i. e. , 90° apart), so that four or more bits are represented by each shift of the combined signal, while the latter combines convolutional codes with QAM.

Unit 10　Text B

1. What is the sampling rate?

The sampling rate is the number of times per second that samples are taken.

2. What is the bit depth?

The bit depth determines the number of possible digital values that can be used to represent each sample.

3. How does a demodulator apply to recover the original signal from the sampled data?

To recover the original signal from the sampled data, a "demodulator" can apply the procedure of modulation in reverse. After each sampling period, the demodulator reads the next value and shifts the output signal to the new value. As a result of these transitions, the signal has a significant amount of high-frequency energy caused by aliasing. To remove these undesirable frequencies and leave the original signal, the demodulator passes the signal through analog filters that suppress energy outside the expected frequency range (greater than the Nyquist frequency). The sampling theorem shows PCM devices can operate without introducing distortions within their designed frequency bands if they provide a sampling frequency twice that of the input signal.

Unit 11　Text A

1. What is WiFi?

WiFi refers to a wireless network, which uses radio waves just like cell phones, televisions and radios do. In fact, communication across a wireless network is a lot like two-way radio communication. Here's what happens:

To start, a computer's wireless adapter translates data into a radio signal and transmits it using an antenna. And then, a wireless router receives the signal and decodes it. The router sends the information to the Internet using a physical, wired Ethernet connection.

2. What does WiFi Hotspots mean?

A WiFi hotspot is simply an area with an accessible wireless network by definition. The term is most often used to refer to wireless networks in public areas like airports and coffee shops. Some are free and some require fees for use, but in either case they can be handy when you are on the go. You can even create your own mobile hotspot using a cell phone or an external device that can connect to a cellular network. And you can always set up a WiFi network at home.

3. How to set up a WiFi connection in your home?

If you already have several computers networked in your home, you can create a wireless network with a wireless access point. If you have several computers that are not networked, or if you want to replace your Ethernet network, you'll need a wireless router. This is a single unit that contains:

- A port to connect to your cable or DSL modem;
- A router;
- An Ethernet hub;
- A firewall;
- A wireless access point.

A wireless router allows you to use wireless signals or Ethernet cables to connect your computers and mobile devices to one another, to a printer and to the Internet. Once you plug in your router, it should start working at its default settings. Most routers let you use a Web interface to change your settings. You can revise the settings according to the instructions. Besides, you can set password to protect your private WiFi.

Unit 11 Text B

1. How can device manufacturers be more confident when bringing new NFC devices to market?

Device manufacturers can now test their products against the latest versions of the Digital Protocol, Tag Type Operations, LLCP, SNEP, and Analog specifications, thus providing added confidence and assurance for companies bringing new NFC devices to market.

2. How does Sony apply NFC?

Sony has built NFC into a growing number of wireless speakers, PCs, smartphones, and other media devices for fast and easy Bluetooth and WiFi pairing.

3. What is the NFC-enabled solution applied in the French company?

A company in France developed an NFC-enabled solution that monitors wine shipment temperatures across distribution channels to ensure the wine's provenance and quality. Each box of wine is equipped with a battery-powered RFID temperature sensor. At each step of the distribution cycle, the wine can be checked for temperature and authenticity using an NFC-enabled device.

Unit 12 Text A

1. In what aspects can DSP be used?

DSP can be used in such aspects as audio and speech signal processing, sonar, radar and other sensor array processing, spectral estimation, statistical signal processing, digital image processing, signal processing for telecommunications, control of systems, biomedical engineering, seismic data processing, among others.

2. What are discretization and quantization?

Discretization means that the signal is divided into equal intervals of time, and each interval is represented by a single measurement of amplitude. Quantization means each amplitude measurement is approximated by a value from a finite set.

3. According to this passage, what does filtering mean? What kinds of filters do you know?

Filtering in this passage refers to a method through which the input signal can be enhanced, namely, the most common processing approach in the time or space domain. You may know the following filters:

1) A "linear" filter is a linear transformation of input samples; other filters are "non-linear". Linear filters satisfy the superposition condition, i. e. if an input is a weighted linear combination of different signals, the output is a similarly weighted linear combination of the corresponding output signals.

2) A "causal" filter uses only previous samples of the input or output signals; while a "non-causal" filter uses future input samples. A non-causal filter can usually be changed into a causal filter by adding a delay to it.

3) A "time-invariant" filter has constant properties over time; other filters such as adaptive filters change in time.

4) A "stable" filter produces an output that converges to a constant value with time, or remains bounded within a finite interval. An "unstable" filter can produce an output that grows without bounds, with bounded or even zero input.

5) A "finite impulse response" (FIR) filter uses only the input signals, while an "infinite impulse response" filter (IIR) uses both the input signal and previous samples of the output signal. FIR filters are always stable, while IIR filters may be unstable.

4. What are commonly used frequency domain transformations?

There are some commonly used frequency domain transformations. For example, the cepstrum converts a signal to the frequency domain through Fourier transform, takes the logarithm, then applies another Fourier transform. This emphasizes the harmonic structure of the original spectrum. Frequency domain analysis is also called spectrum or spectral analysis.

Unit 12 Text B

1. What is the advantage of a specialized digital signal processor?

A specialized digital signal processor, however, will tend to provide a lower-cost solution, with better performance, lower latency, and no requirements for specialized cooling or large batteries.

2. Do DSPs support virtual memory? Why?

DSPs frequently use multi-tasking operating systems, but have no support for virtual memory or memory protection. Operating systems that use virtual memory require more time for context switching among processes, which increases latency.

3. Can modern signal processors perform well? Why?

Modern signal processors yield greater performance; this is due in part to both technological and architectural advancements like lower design rules, fast-access two-level cache, (E) DMA circuitry and a wider bus system.